Surveying Instruments and Technology

Surveying Instruments and Technology

Leonid Nadolinets, Eugene Levin, and Daulet Akhmedov

CRC Press is an imprint of the
Taylor & Francis Group, an **informa** business

CRC Press
Taylor & Francis Group
6000 Broken Sound Parkway NW, Suite 300
Boca Raton, FL 33487-2742

Printed on acid-free paper

International Standard Book Number-13: 978-1-4987-6238-0 (Hardback)

Library of Congress Cataloging-in-Publication Data

Names: Nadolinets, Leonid, author. | Levin, Eugene, author. | Akhmedov, Daulet, author.
Title: Surveying instruments and technology / Leonid Nadolinets, Eugene Levin, and Daulet Akhmedov.
Description: Boca Raton : Taylor & Francis, CRC Press, 2017. | Includes bibliographical references and index.
Identifiers: LCCN 2016058868 | ISBN 9781498762380 (hardcopy : alk. paper)
Subjects: LCSH: Surveying--Instruments. | Geospatial data. | Geodesy.
Classification: LCC TA562 .N33 2017 | DDC 681/.76--dc23
LC record available at https://lccn.loc.gov/2016058868

Visit the Taylor & Francis Web site at
http://www.taylorandfrancis.com

and the CRC Press Web site at
http://www.crcpress.com

Contents

Preface

Surveying is one of the most ancient professions, and in recent years it has been boosted by technology advances. These advances come from developments in optical, mechanical, electronics, aerospace, sensor, and information technologies. To this end, instrumentation deployed by modern surveyors is associated with a significant body of knowledge. This book is about surveying instruments and covers all aspects of them, including historical references, physical and constructional principles of operation, and features of modern instruments. The idea of creating a book about surveying instruments occurred to us while speaking with our colleagues and customers. Existing books in the field of surveyor's instruments were published many years ago and do not include up-to-date surveying instruments information. We believe this book is long overdue as a complete description of the latest surveying instruments.

This book is suitable for undergraduate and graduate students who are taking surveying courses, and professional surveyors seeking a better understanding of surveying instruments. Field surveyors working on large construction projects, such as road networks and buildings, and other geodetic control networks may find useful details presented in this book. Surveying business owners may also benefit.

It is hoped that this book will help readers to become aware of technological solutions behind the development of surveying instruments.

Acknowledgment

The authors take this opportunity and express a cordial gratitude to Trimble for support of the current book preparation by sharing with us materials on the UX5 HP Unmanned Aircraft system, which were very helpful with the unmanned aerial vehicle discussion.

We are very thankful for the help that came from Elena Sviridova, spouse of author Leonid Nadolinets. Elena was not only patient during our work, but also participated very actively in book writing and editing.

We are also very pleased to express our gratitude to Jeremiah Harrington of the Michigan Tech Research Institute (MTRI) for his help.

Very special thanks goes to Almat Raskaliyev, a PhD student in the Department of Mechanics, Faculty of Mechanics and Mathematics of Al-Farabi Kazakh National University. He assisted in the preparation of the global navigation satellite system (GNSS) chapter.

Authors

Leonid Nadolinets lives in Moscow, Russia. Nadolinets has devoted his life to studying and developing surveying instruments. His PhD thesis covered topical surveying instrument developing. Nadolinets has worked as a research fellow at many research institutes in Russia and abroad. He has also attended training courses arranged by leading surveying manufacturers from the United States, Europe, Japan, and China. Currently, Nadolinets heads a consulting company that develops and maintains surveying instruments.

Eugene Levin is the program chair of surveying engineering and associate professor at the School of Technology at Michigan Technological University. Levin also directs the Integrated Geospatial Technology graduate program. He earned an MS degree in astrogeodesy from Siberian State Geodetic Academy in 1982 and a PhD in photogrammetry from Moscow State Land Organization University in 1989. He has more than 30 years of experience in academia and the geospatial industry in the United States, Israel, and Russia. He has held research and management positions with several academic institutions and high-tech companies, including the Research Institute of Applied Geodesy, Omsk Agricultural Academy, Rosnitc "Land," Ness Technologies, Physical Optics Corporation, Digital Map Products, American GNC, and Future Concepts. He has served as a principal investigator and project manager for multiple award-winning government programs.

Daulet Akhmedov has a PhD in theory of mechanisms and machines, and a PhD+ in geo-information and geotechnology. He is currently a director at the Institute of Space Technique and Technology, in Almaty, Kazakhstan. His research interests include mathematical models and numerical methods for solving the problem of simulating motion of mine dumps of various models, mainline, and shunting locomotives and trains; mathematical models and numerical methods for high-precision satellite navigation; design and development of geographic information systems (GIS); design, manufacture, and implementation of communication systems based on low-orbit satellite communication systems; design, manufacture, and implementation of high-precision satellite navigation systems; and development of dispatching systems based on the use of low-orbiting satellite systems, VHF and GSM communications, satellite navigation technologies, and GIS technologies.

1 Introduction

1.1 SURVEYING PROFESSION AND INSTRUMENTS: HISTORICAL PERSPECTIVE

The most ancient of known maps were found in Spain. They were engraved 13,667 years ago on a hand-sized rock and probably were made by Magdalenian hunter-gatherers (Bates 2009) (see Figure 1.1). In spite of ease of use, the metric quality of that map is questionable and it was produced based on the visual perception and imagination of the hunter-gatherers. Thus we may call them the first cartographers, but not surveyors. The surveying profession combines professionalism in measurements on terrain with social (land-laws related) responsibilities.

The first historically documented professional surveying responsibilities can be attributed to the famous thumb of Menna dated 1400 BC. Menna lived during the New Kingdom of Egypt (Eighteenth Dynasty), 3500 years ago. He was one of the surveying crew that consisted of four royal scribe surveyors (the others being Amenhotpe-si-se, Djeserkareseneb, and Khaemhat), through whose funerary monumentation can be seen in Figure 1.2 ("Menna–TT69," n.d.). Distance measurement instruments of that time were called "snakes" and basically comprised the same principles as the measurement tapes used by surveyors until the middle of the twentieth century (Figure 1.3). Steel tape measures are used for home property survey, property line survey, and any other survey that requires a distance measurement. Survey crews still use steel tape measures, even in the era of high tech electronic survey equipment. As for angular measurements, a pioneering instrument was developed by Eratosphenes (Figure 1.4), who used the first sun-ascending angle instrument, called the gnomon (sundial), for accurate definition of the circumference of the Earth around 240 BC (Smith 2005).

Ancient Romans also contributed to surveying instrument development by creating a groma (known as Roman measurement cross) by Heron of Alexandria (O'Connor and Robertson 1999) (Figure 1.5) and the first level prototypes by Vitruvius (Opdenberg 2008). The dioptra (diopter), invented in Greece 300 BC as an astronomic instrument (Figure 1.6), was used by astronomers and then adopted by surveyors. The dioptra was a sighting tube, or alternatively a rod with a sight at both ends, attached to a stand. If fitted with protractors, it could be used to measure angles and can be considered as a prototype of modern theodolites (Evans 1998).

In the sixteenth century, Leonardo da Vinci invented a mechanical odometer designed to measure the distance traveled by a carriage (Moon 2007) (Figure 1.7).

Significant progress in surveying instrument development was achieved due to invention of the telescope by Galileo (Van Helden 2004). In 1593 German mathematician Cladius developed the nonius principle (Berry 1910), which is widely deployed in modern surveying instruments (Figure 1.8).

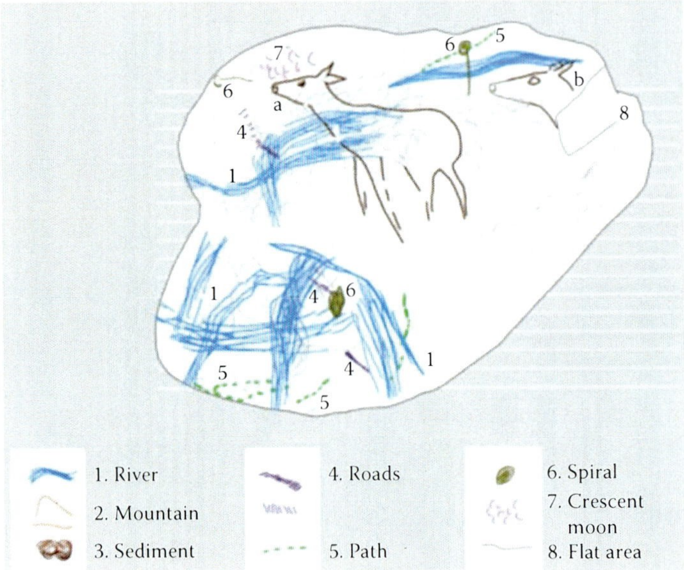

FIGURE 1.1 Ancient map.

In 1662 French engineer Thevenoux invented the cylindrical level (Deumlich 1982) and in 1674 Italian scientist Mantanari designed a telescope with parallel wires to measure distances ("Register: Milestones of geodesy," n.d.). British scientists Sisson and Ramsden, in the eighteenth century, developed the first theodolite, principally almost analogous to modern instruments (McConnell 2013). The start of the electronic distance measurer (EDM) can be attributed to invention in Sweden by Bergstrand of the phase measurements principle realized in the device later called the geodimeter (Bergstrand 1960). The theodolite and geodimeter are depicted in Figure 1.9.

GPS and unmanned aerial vehicles (UAVs) are relatively new technologies deployed by surveying engineering at the end of twentieth and beginning of the twenty-first centuries. Therefore, a short history of these technologies is reviewed in respective chapters of this book.

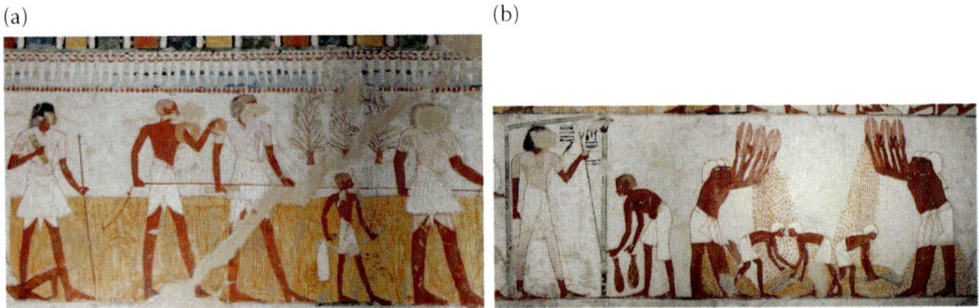

FIGURE 1.2 Scribe Djeserkareseneb carrying out a survey of the crops (a) and Scribe Menna looks authoritatively over his surveying party (b). (© Osirisnet.net)

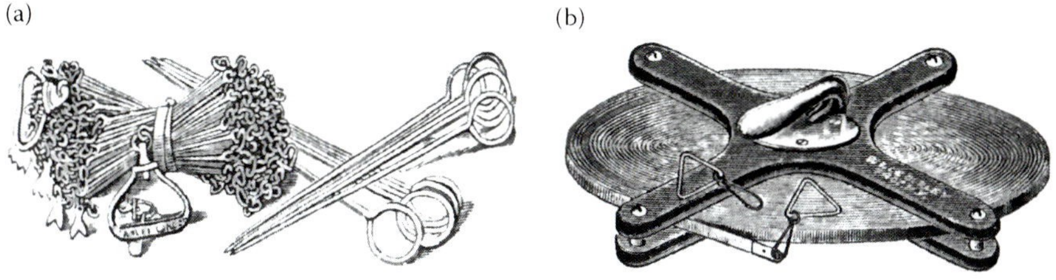

FIGURE 1.3 Metallic tape was used as the main distance-measuring instrument until the first half of the twentieth century (a). Surveying distance measurements with a tape (b). (Adapted from Heanders. 2012. Using a steel tape measure to survey. InfoBarrel. http://www.infobarrel.com/Using_A_Steel_Tape_Measure_To_Survey.)

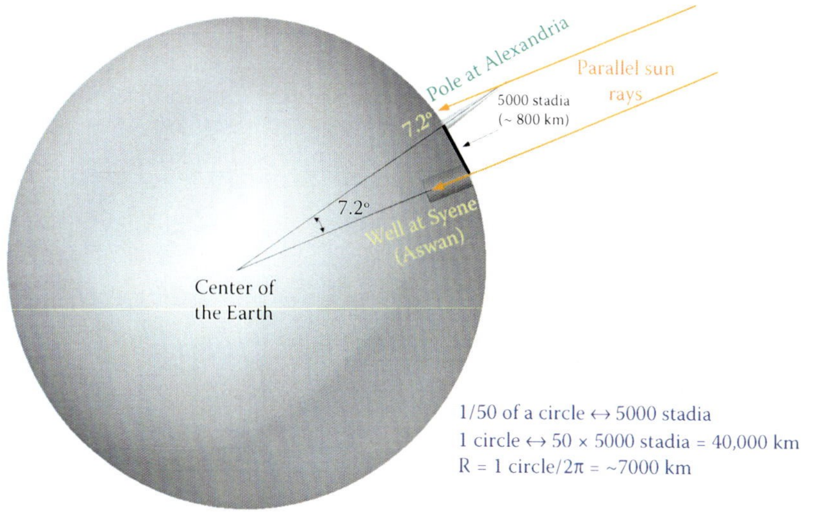

FIGURE 1.4 Principles of Earth circumference measurement by gnomon-sundial (Eratosphenes project).

Nowadays developing integrated systems are being developed to allow collection of vast amounts of 3D data deploying lidar scanner integration. There are multiple publications containing reviews of lidar instruments and technology, therefore we leave it out of scope of this book.

1.2 EXISTING BOOKS AND WHY WE DECIDED TO WRITE OUR BOOK

We researched existing texts on surveying instruments. We recommend the following texts (listed in chronological order):

1. *A Manual of the Principal Instruments Used in American Engineering and Surveying*, by W. & L. E. Gurley, 1871

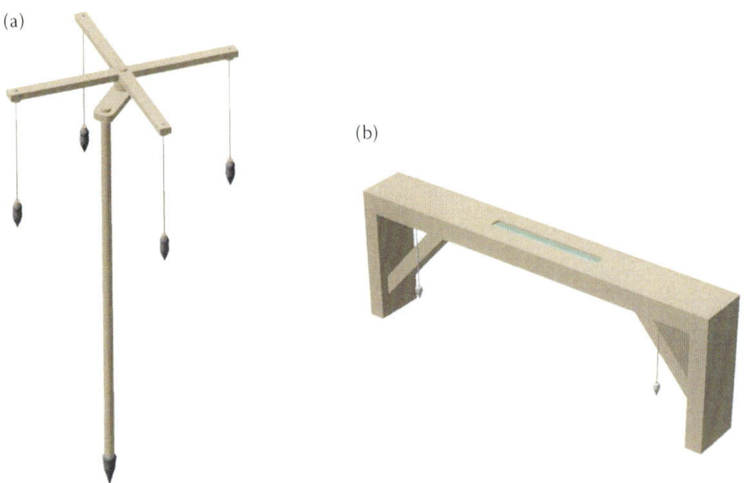

FIGURE 1.5 Ancient Roman surveying instruments groma (a) and chorobates (b).

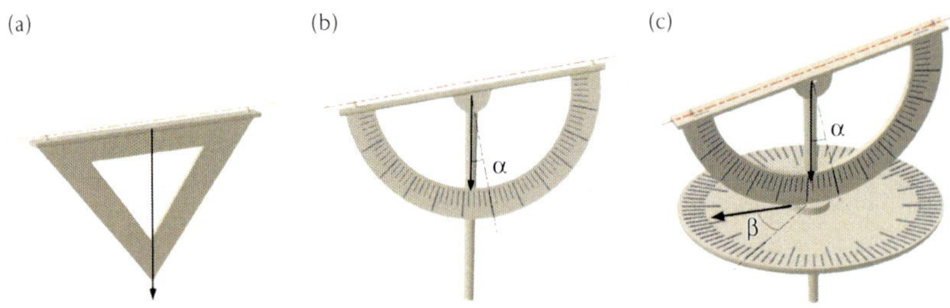

FIGURE 1.6 Three stages in the evolution of the dioptra, from a simple horizontal sight to a multi-triangle measuring tool.

FIGURE 1.7 Da Vinci odometer.

2. *A Manual of Modern Surveying Instruments, and Their Uses; Together with a Catalogue and Price List of Scientific Instruments*, by A. Lietz Company, 1911–12
3. *Surveying and Levelling Instruments*, by W. F. Stanley, 1914
4. *Principles and Use of Surveying Instruments*, by James Clendinning, 1972

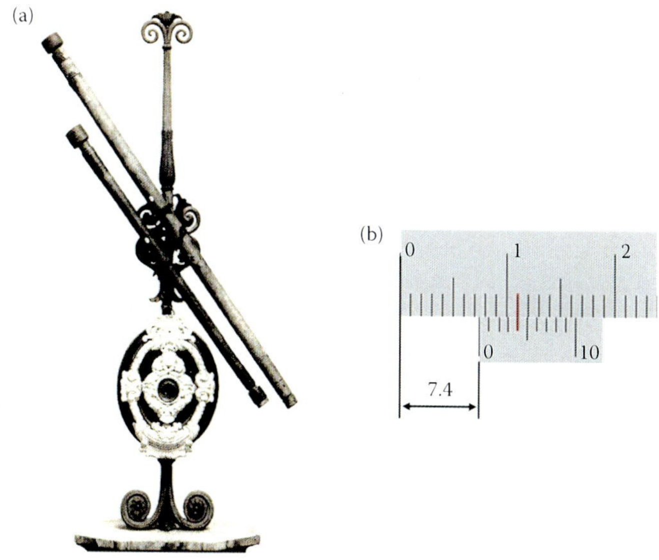

FIGURE 1.8 Telescope of Galileo (a) and nonius principle (b).

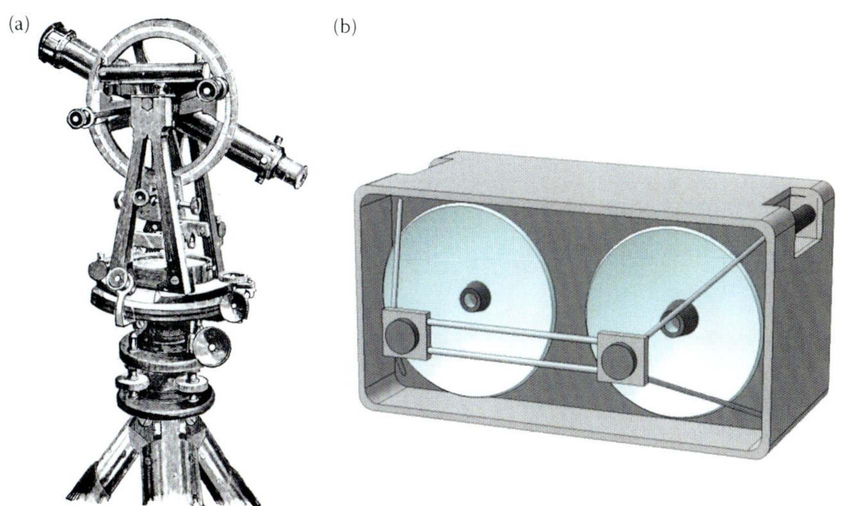

FIGURE 1.9 The first theodolite by Sisson and Ramsden (a), and the first EDM, the geodimeter (b).

5. *Surveying Instruments*, by Fritz Deumlich, 1982
6. *Surveying Instruments and Their Operational Principles*, by L. Fialovszky, 1991
7. *Electronic Distance Measurement: An Introduction*, by Jean M. Rüeger, 1996
8. *Surveying Instruments of Greece and Rome*, by M. J. T. Lewis, 2001
9. *Instrumentenkunde der Vermessungstechnik*, by Fritz Deumlich and Rudolf Staiger, 2001
10. *Methodologie voor de nauwkeurigheidscertificatie van totaalstations met ondersteuning voor satellietplaatsbepaling*, by Alain Muls and Alain De Wulf, 2007
11. *Engineering Surveying*, 6th ed., by W. Schofield and Mark Breach, 2007
12. *Elementary Surveying: An Introduction to Geomatics*, 13th ed., by Charles D. Ghilani and Paul R. Wolf, 2012

We especially recommend the first three titles for those who are interested in the history of surveying instruments. Illustrations in those books are works of art. Good and detailed chapters of surveying engineering instruments are presented in books 8 and 9. Books 4 to 8 give a good perspective of surveying instrument development of the 1950s to 1980s. Book 11 is an up-to-date engineering surveying textbook describing applications of surveying knowledge in practice. There are also sections about surveying instruments, including levels, EDMs, and theodolites. Book 12 is a fundamental surveying engineering textbook widely used in the United States for surveying workforce preparation. It has excellent coverage of surveying basics and reviews some information about surveying instruments, including sections about levels and EDMs.

Our major motivation to write this book was to cover gaps in the existing surveying engineering literature by reviewing, on a very detailed level, the most current innovations in surveying instrument technology along with detailed specification on state-of-the-art surveying engineering equipment.

1.3 STRUCTURE OF THIS BOOK

This book chapters are arranged based on surveying instrument functionality. Every chapter discusses physical and mathematical measurement principles along with state-of-the-art technology samples of realization of those principles.

Chapter 2 is devoted to description of levels and covers principles from the classic bubble-based mechanical levels to the most advanced microelectromechanical systems (MEMS) technology, which enables automatic levels including modern laser plane systems.

Chapter 3 describes theodolites and their parts, including a review of current technical capabilities and technology realized in the most advanced theodolite systems.

Chapter 4 covers major principles and technologies associated with electronic distanced measurements. It contains the mathematical basis of pulse- and phase-based methods and describes practical realizations of these methods by state-of-the-art EDM providers.

Chapter 5 has a description of total stations, including modern robotic systems with samples of technologies deployed and state-of-the-art technical characteristics.

Chapter 6 covers global navigation satellite system (GNSS) instrumentation principles and realization samples. It has a detailed description of GNSS/GPS components and functionality. It also has a list of the state-of-the-art system technical specifications.

Chapter 7 is about surveying engineering application of unmanned aerial vehicles (UAVs) and describes the sensors and platforms along with principles of UAV imagery processing for surveying engineering applications. This chapter also has a detailed discussion on practical planning and execution of small UAV surveying missions, including both hobbyist and professional-level platforms.

REFERENCES

Bates, C. 2009. Oldest map in western Europe found engraved on 14,000-year-old chunk of rock. *Daily Mail*, August 6. http://www.dailymail.co.uk/sciencetech/article-1204539/Oldest-map-western-Europe-engraved-14-000-year-old-chunk-rock.html.

Bergstrand, E. 1960. The geodimeter system: A short discussion of its principal function and future development. *Journal of Geophysical Research* 65(2):404–409.

Berry, A. 1910. *A Short History of Astronomy*. London: Scribner.

Deumlich, F. 1982. *Surveying Instruments*. Berlin: Walter de Gruyter.

Evans, J. 1998. *The History and Practice of Ancient Astronomy*, pp. 34–35. New York: Oxford University Press.

Heanders. 2012. Using a steel tape measure to survey. InfoBarrel. http://www.infobarrel.com/Using_A_Steel_Tape_Measure_To_Survey.

McConnell, A. 2013. Jesse Ramsden: The craftsman who believed that big was beautiful. *The Antiquarian Astronomer* 7:41–53.

Menna–TT69. n.d. Osirisnet.net. http://www.osirisnet.net/tombes/nobles/menna69/e_menna_01.htm.

Moon, F. C. 2007. *The Machines of Leonardo Da Vinci and Franz Reuleaux: Kinematics of Machines from the Renaissance to the 20th Century*, Vol. 2. Netherlands: Springer Science & Business Media.

O'Connor, J. J. and E. F. Robertson. 1999. Heron of Alexandria. *The MacTutor History of Mathematics Archive*. http://www-history.mcs.st-andrews.ac.uk/history/Biographies/Heron.html.

Opdenberg, G. 2008. The chorobates of Vitruvius evaluated by a surveyor. *Archaologisches Korrespondenzblatt* 38(2):233–246.

Register: Milestones of geodesy. n.d. http://www.in-dubio-pro-geo.de/?file=register/mstones&english=1.

Smith, W., ed. 2005. Eratosthenes. In *A Dictionary of Greek and Roman Biography and Mythology*. University of Michigan Library, Ann Arbor, MI.

Van Helden, A. 2004. Galileo's telescope. *MyScienceWork*. https://www.mysciencework.com/publication/show/beb376bfd7abca0ef53e30c45a08ec52.

2 Levels

The cornerstone of any surveyor's leveling instrument is a device that creates a horizontal line, or a surface perpendicular to a plumb line. Ancient pendulum and hydraulic leveling systems have been useful for several thousand years. However, those systems have reached the limits of their technological accuracy.

In the second half of the seventeenth century a tubular level was invented. It is considered a significant technological breakthrough. The tubular level was based on a simple principle, and it put into motion the advanced technologies of that age, namely, skills in accurate grinding of glass surfaces. A tubular level is a glass tube filled with fluid and sealed on both ends (Figure 2.1). The internal surface of the vial is arc-shaped. Spirits or other easy-flowing liquids are used to fill the vial. First, the vial is filled up with hot liquid, and then after the liquid cools a vapor bubble is formed. The external surface of the tube has a scale graduated at 2 mm intervals (up-to-date standard). The vial is fitted with a protective metal shield with lugs for fixing and adjusting screws. The operating principles are simple and did not demand any explanations even in the seventeenth century. Nowadays, tubular levels in surveying instruments are applied with graduated scale divisions from 1′ to 10″. The vial radiuses are from 6.9 to 41.2 m, correspondingly. Polishing precision of the vial is such that the height difference for a 10″ arc is only 0.1 μm. When touching the vial surface with your finger, slight movement of the bubble can be seen. Avoid any local warming of the vial. Earlier, a tubular level was used with diopters, known from Ancient Greece. This instrument was named in *Cyclopedia* (published by Ephraim Chambers in London in 1728) as an "air level." Also a method of leveling with the aid of two rods was described there. That practice is no different from that of today.

A diopter level consisted of a frame with a tubular level in the central part, and two diopters were fixed on the edges (Figure 2.2). The frame could be inclined with the aid of the leveling screw to set the bubble in the central position, and then sighting was carried out by the means of the horizontal diopters. One of the diopters could be adjusted at height. It was necessary to keep the leveling basic rule: *The sighting line must be parallel to the tubular level axis.*

Let us determine the instruments accuracy on the assumption of the following parameters: angular accuracy of diopter sighting is about 30″, the tubular level scale interval is 20″, and the distance from the instrument to the rods is 10 m (no use watching longer distances with an unaided eye). It turns out that it was possible to measure elevation within 2–3 mm accuracy!

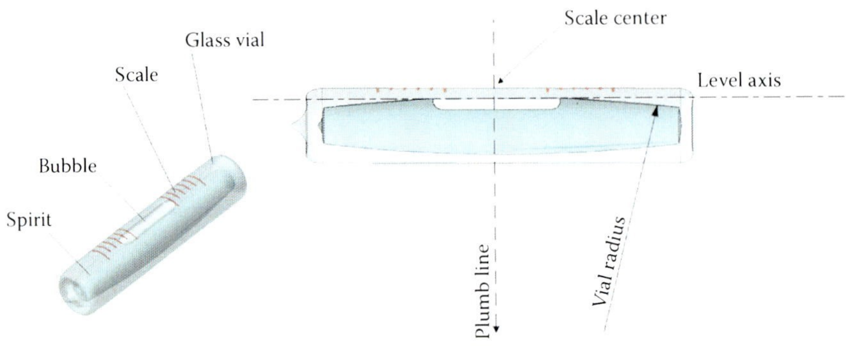

FIGURE 2.1 Tubular level.

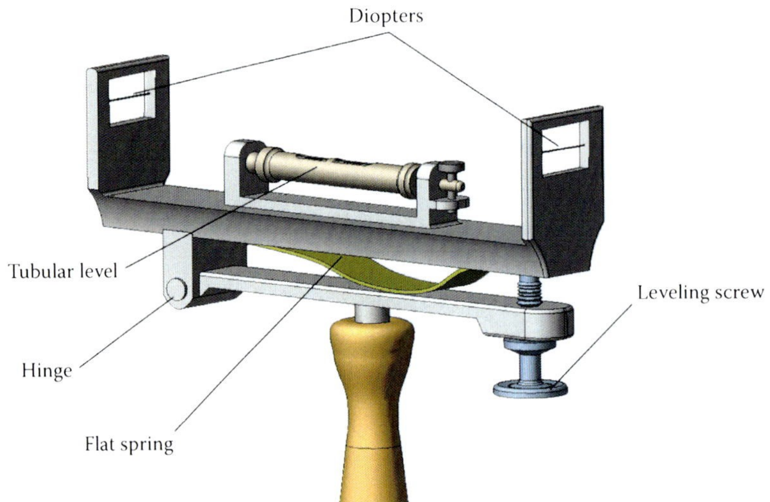

FIGURE 2.2 Diopter level.

2.1 OPTICAL LEVELS

2.1.1 Dumpy Levels

Invented by Johannes Kepler at the beginning of the seventeenth century, an optic tube known as the Kepler telescope (Figure 2.3) was another significant step in the development of surveying instruments.

The telescope consisted of only two components. These were two positive lenses with different focuses. While watching infinitely distant points, stars, for example, parallel rays are collected in the objective lens focus. The ocular front focus was set in the objective focus. A reticle would be placed at this point. In the past, the reticle could not be set in Galileo's telescope because a negative lens was applied in the eyepiece. The negative lens does not have the actual focus point. Kepler telescope magnification is the ratio of the objective and ocular focuses. The telescope produces a reverse image.

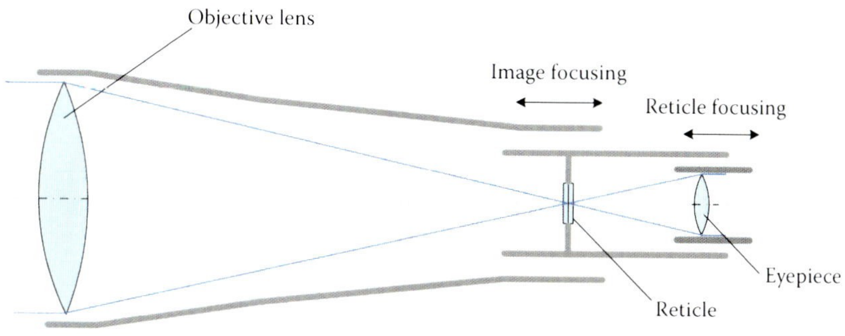

FIGURE 2.3 Kepler telescope.

Using the Kepler telescope instead of diopters was another logical step in the advancement of surveying instruments. In 1703, almost a hundred years after the appearance of the Kepler telescope, Alain Manesson Mallet constructed a level known as a dumpy level (see Figure 2.4). The concept of this device has prevailed for over two and a half centuries. Even nowadays, instruments are made based on this principle.

The model of a typical dumpy level from the end of the nineteenth century is shown in Figure 2.4. It consisted of a Kepler telescope with external focusing, a tubular level fastened to it with the aid of adjusting screws, a frame on a rotating foundation, and foot screws to incline the foundation. The bubble movement to the center of a tubular level was achieved by the means of the foot screws, and a reticle horizontal setting of the telescope was carried out. Some instruments were equipped with an additional circular level (see Figure 2.5) or a less accurate tubular level, which simplified preliminary adjustment of the instrument. Also some devices had aiming screws to improve targeting.

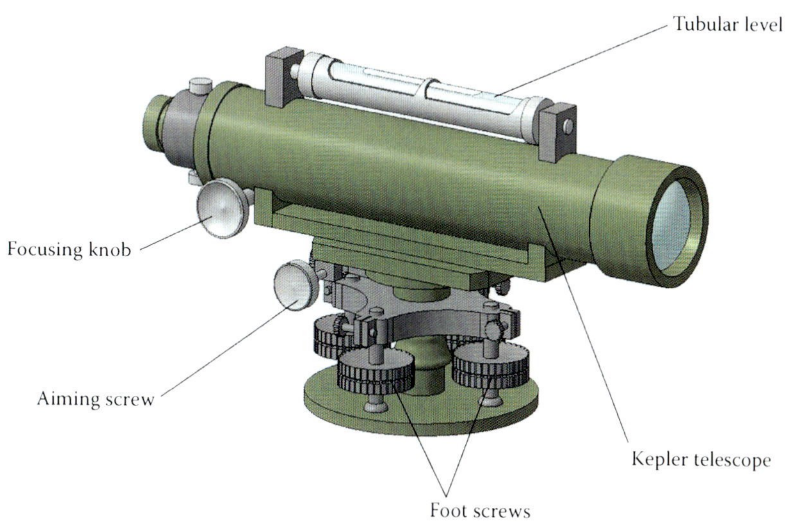

FIGURE 2.4 Dumpy level.

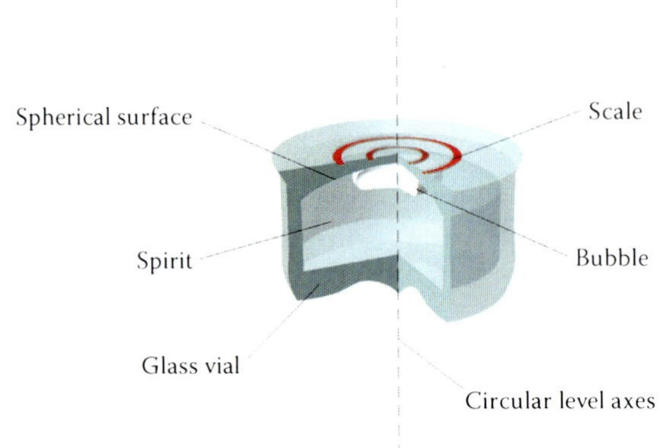

FIGURE 2.5 Circular level.

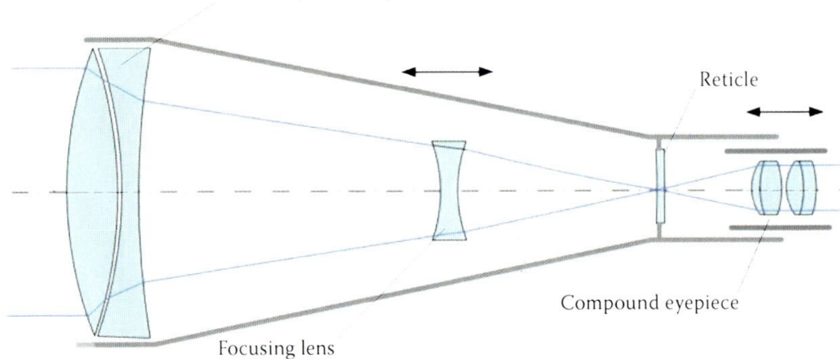

FIGURE 2.6 Kepler telescope with an internal focusing.

The appearance of the Kepler telescope with an internal focusing lens (see Figure 2.6) was the next step in dumpy level development.

Thanks to the fact that the size of the instrument was diminished and it became more hermetic. With time the main optic components of the telescope have become more complicated. A three-lens system was applied as an objective, called a teleobjective. Spherical and chromatic aberrations were corrected in it. Also, ocular optics started becoming more sophisticated. Made of two or three components, it was free of geometric distortions and had a wide field of view with acceptable magnification.

The direct image realization of the Kepler telescope was a final step toward modern devices. There are two ways of inverting of an image. It can be fulfilled by the means of an inverting eyepiece or a system of prisms. The first way demands a multicomponent eyepiece and lengthens the telescope. In the second case, prisms are applied, and the length of the telescope shortens. Therefore, the prism systems are more often used in surveyor's devices.

An optical prism is a transparent element with one or several reflecting flat surfaces. They can be covered with mirrored metal coverings or use a full internal reflecting effect.

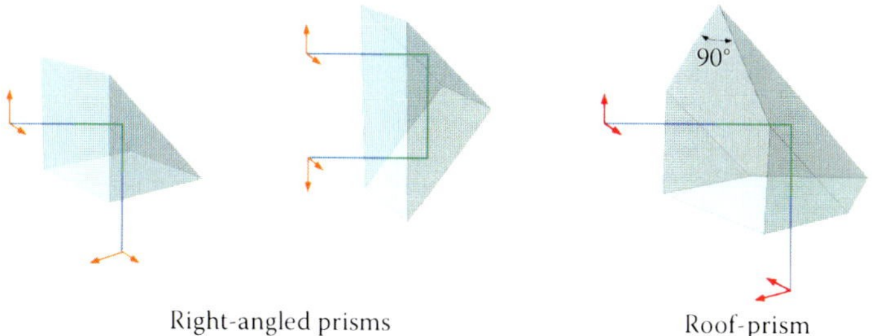

FIGURE 2.7 Right-angled triangular prisms.

An elementary prism works as a mirror and rotates the image only around one axis (see Figure 2.7). A full turn of the image is fulfilled by so-called roof prisms which have two, instead of one, reflecting surfaces at a right angle to each other. These two types of prisms are the basis for construction of more complicated prismatic systems. A double Porro prism consists of two elementary prisms placed at right angles to each other. This solution allows a full reversal of the image. Direct and reversed images have the same direction and are parallel to each other (the left side of Figure 2.8). This prism system is very often applied in binoculars.

Porro-Abbe prisms consist of three elementary prisms stuck together (the right side of Figure 2.8). This system also creates a completely reversed image that coincides with an incoming direction. This system is often used in the telescopes of modern surveyor instruments.

Abbe-Koenig prisms are assembled as two prisms stuck together at a 120° angle (see Figure 2.9). One of their three reflecting surfaces should be a roof prism. This system of prisms fulfills a full reverse of the image and has a very important feature: the direct and reversed images' directions and also their positions are not changed. Because of this feature, the prism is widespread in telescopes of geodetic tools and particularly in dumpy levels.

Figure 2.10 shows how an up-to-date dumpy level is arranged. A dumpy level telescope consists of a three-component objective: a teleobjective, an inverting prism, and an eyepiece. A frontal lens of the objective is often used as a duplex achromat. The middle lens of the teleobjective is mobile (focusing). As a rule, the Abbe-Koenig

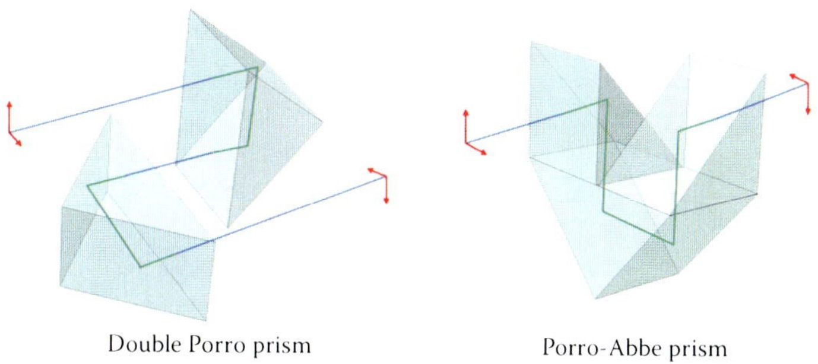

FIGURE 2.8 Double Porro and Porro-Abbe prisms.

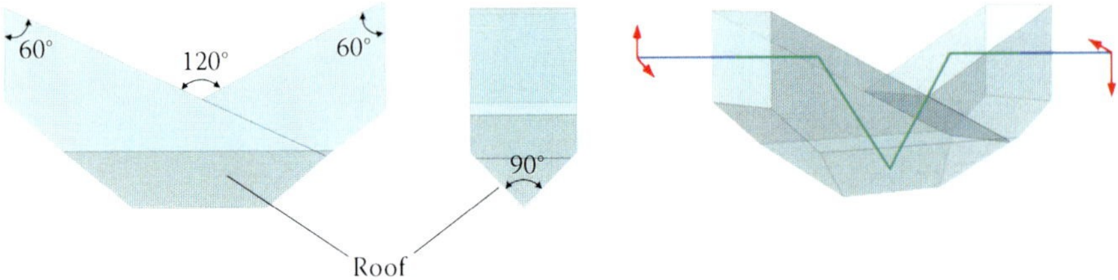

FIGURE 2.9 Inverting Abbe-Koenig prism.

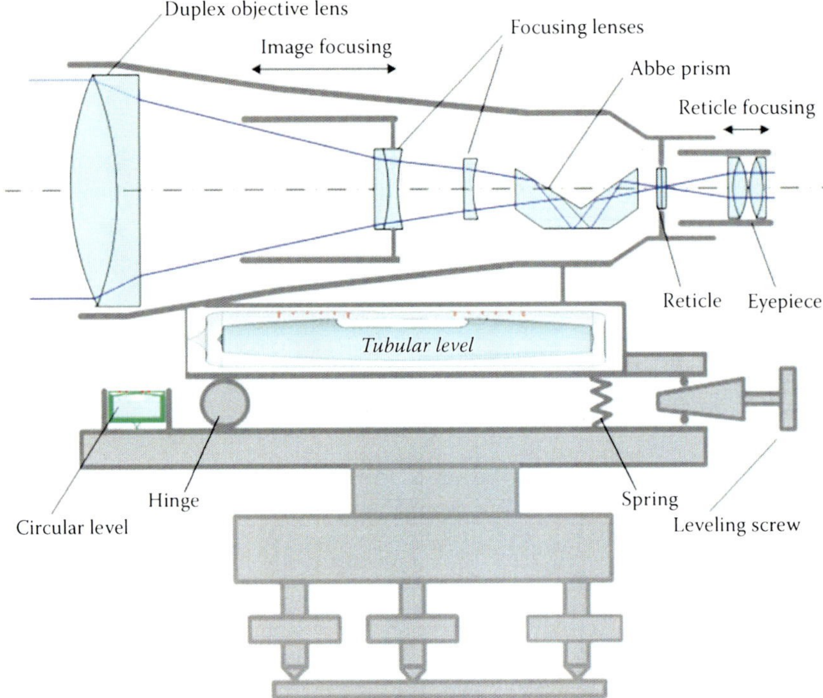

FIGURE 2.10 Up-to-date dumpy level.

style prism is more often used in levels to reverse the image. The reticle has a protective glass cover for protection against dust.

Dumpy level-type telescopes have a magnification from 20× to 40×, and are able to focus from 0.8 m to infinity. Tubular levels that have 10″ to 30″ sensitivity per 2 mm are applied in dumpy levels. The telescope and the tubular level have a common base where the level is fitted on by the means of an adjusting joint. There are adjusting screws to keep the main rule for a surveyor's dumpy level. This rule is the following: *The collimation axis of the telescope and the level axis should be parallel. The collimation axis is the line passing through the objective optical center and the reticle center.*

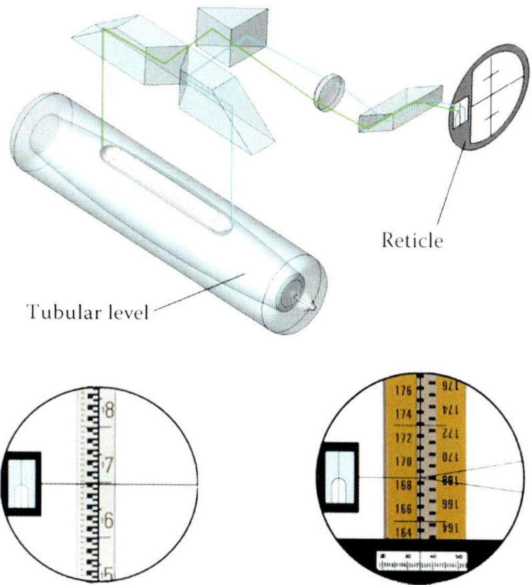

FIGURE 2.11 Visual field of a dumpy level.

The telescope and the tubular level are joined with the rotating base by the aid of a leveling mechanism. It consists of a hinge, a reflexive spring, and a leveling screw. The axis of the rotating base is inserted into the tribrach collar. Three-foot screws are necessary for a preliminary leveling of the rotating base by the means of the circular level that it sets on.

Modern dumpy levels have various appliances to supervise the tubular level bubble position. The elementary of them is a clapper-type mirror. Prismatic systems, which transfer the bubble edges in eyepiece's field of sight, are more complicated (see the top part of Figure 2.11).

The field of sight for a moderate accuracy dumpy level is in the left lower part of Figure 2.11. Here, we also see a standard leveling reticle. It consists of cross vertical and horizontal lines. Still, there are two horizontal lines for distance measuring.

High precision levels have a more complicated field of view (the right lower side of Figure 2.11). A divergent bisector is used instead of the right part of the horizontal line of the reticle. It is convenient to superpose with divisions on a precise leveling rod. High precision levels have a micrometer scale image that is transferred into the sighting field. This scale is mechanically connected with a driver from a plane-parallel plate. This glass plate is placed before the dumpy level objective.

Before reading-out the leveling rod it is necessary to set the bubble of the tubular level in the central position by the means of the leveling screw. The measurements fulfilled by the means of a dumpy level are very tiring and out of date. So why are levels of this type still used? The following sections explain some reasons for it.

2.1.1.1 Accuracy

High-precision dumpy levels can have a standard deviation for 1 km double-run leveling of 0.2 mm. Such devices are the PL1-31(Sokkia), TS-E1 (Topcon), and

TABLE 2.1
Modern Dumpy Levels

Model	Accuracy (mm/km)	Magnification ($n\times$)	Accuracy of Main Level (n''/2 mm)	Minimal Focusing Range (m)	Manufacturer
PL1-31	0.2	42	10	2.3	Sokkia
NABON	0.2	42	8	0.3	Breinhaupt-Kassel
TS-E1	0.2	42	10	2	Topcon
TS-3A	1.5	32	40	1.4	Topcon
TS-3B	2	26	120	0.5	Topcon
NI-3	2.5	30	15	1.3	IPZ-UA
DS3-D	3	32	22	2	Geo-Master
3N-5L	5	20	30	1.2	UONZ

NABON (F.W. Breithaupt & Sohn GmbH & Co. KG). Some high precision automatic levels have accuracy only from 0.3 to 0.8 mm. It is because of this fact that we cannot know definitely the plumb line direction. This uncertainty affects a dumpy level measurement results only once; meanwhile the uncertainty is present in an automatic level at least twice. Automatic levels have an optical reflecting surface that is connected with a pendulum. The reflecting surface doubles the deflection of an optical beam (e.g., the mirror is set at an angle of 45° to turn up the beam at the right angle). This explains why this uncertainty of the plumb line direction worsens the accuracy of automatic levels.

2.1.1.2 Reliability

Dumpy levels have better shock resistance than automatic levels. Automatic levels have some optical details suspended on thin tapes. After the same impact, a dumpy level may demand user adjustment, whereas an automatic level could require a repair because of breakage or deformation of the spring ribbons on the compensator.

2.1.1.3 Insensitivity to Strong Magnetic and Electromagnetic Fields

Automatic levels compensators have magnetic or air dampers. The influence of electromagnetic industrial fields and the Earth's magnetic field may affect the magnetic damper levels.

Various types of dumpy levels are shown in Table 2.1.

2.1.2 Automatic Levels

2.1.2.1 Automatic Optical Levels

Automatic levels appeared much earlier than dumpy levels. Well known since ancient times, a dioptical level designed with a pendulum could be considered the earliest automatic level. When telescopes appeared, people also tried to fix it at a pendulum. The result was unsatisfying because a telescope's center of gravity changes while refocusing. Any attempts to put a focusing element vertically did not reach the accuracy of dumpy levels. Step by step people realized that they did not need to install the whole telescope at a pendulum. It was better to fix one or several optical components at the pendulum. There were several ways to solve the problem:

- Place a compensating element before the objective lens of the telescope
- Use a moving objective lens
- Place a compensating element between the objective lens and a focusing component
- Implement an inclining focusing component
- Use a moving part of an inverting image component
- Use a moving reticle

The unit applied to solve these problems is called a compensator. The compensator operates in accordance to the main rule of an automatic level: *The collimation axis of the telescope should be perpendicular to the plumb line in the whole operating range of the compensator.*

Quick development of automatic levels began approximately in the 1930s, and the "Golden Age" of automatic levels lasted until the 1990s. During the last twenty years, the compensator has been used as an image-inverting unit in 90% of automatic levels. In this case the compensator is set between the focusing lens and the reticle (see Figure 2.12). Let us imagine that the automatic level telescope would be inclined at corner α. In order to make the image come back to the same place the compensator should incline the image at corner β, which is a little more than corner α. The direction of corner β should be opposite to the direction of the telescope's inclination since the compensator rotates the image. Obviously, a usual pendulum cannot be applied to solve this problem. In this case, we can apply a special pendulum solution to transform the angle inclination. There are two types of this solution: an inverse pendulum or a pendulum added with a suspension made of crossed flexible tapes (Figure 2.13).

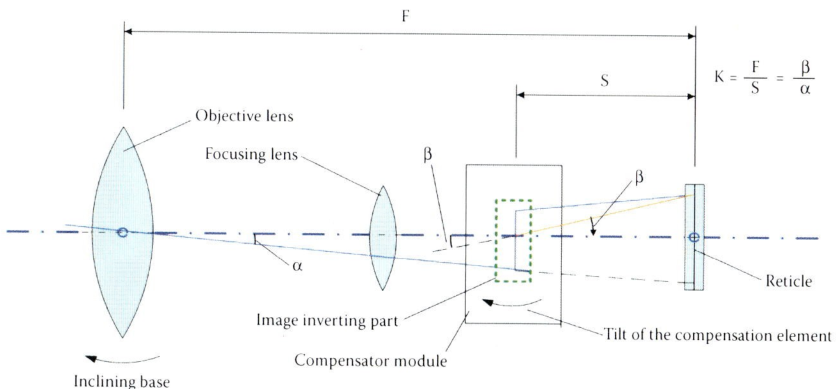

FIGURE 2.12 Operating principle of a compensator.

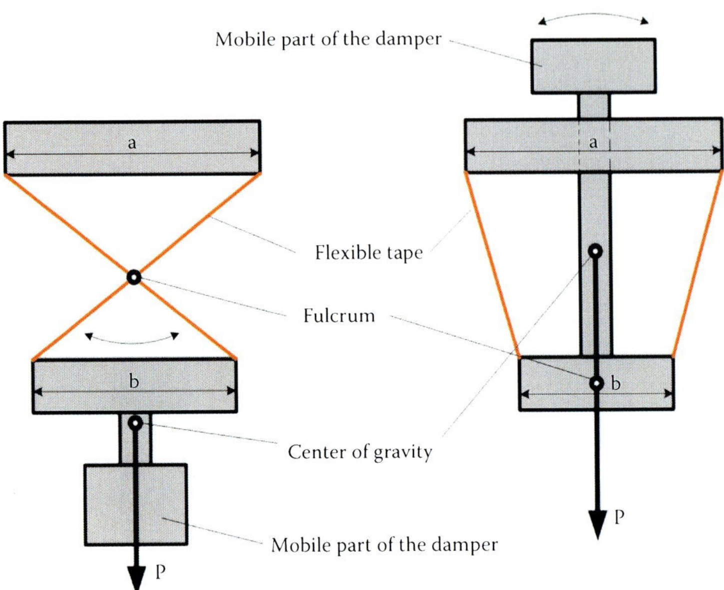

FIGURE 2.13 Kinematic scheme of a compensator.

The value of angle transformation is called compensation factor K:

$$K = \frac{F}{S} = \frac{\beta}{\alpha} \tag{2.1}$$

F is the focal distance of automatic level teleobjective.

There are several ways to affect the compensation factor K:

1. Changing the distance S from the compensator to the reticle. The compensator sensitivity increases depending on the distance to a reticle.
2. Choosing the type of a suspension for the compensating element. Let us define Km as the compensation factor, which depends on suspension bracket features. Different kinds of suspension brackets in the compensator, and how the Km depends on them are shown in Figure 2.14. Static and mobile parts of the compensator are marked, respectively, a and b.
3. Choosing the compensator optical scheme. In this case, it is necessary to know the way of the optical beams in the compensator and the quantity of the reflecting surfaces in their optical scheme.
4. We should take into account the quantity of operating reflecting surfaces only in the movable optical elements of the compensator. The reflecting surface doubles deflection of an optical beam. As a result the optical compensation factor Ko cannot be less than 2. Most modern levels compensators are constructed according to the optical scheme of inverting Abbe-Koenig or Porro-Abbe systems, where some elements are mobile (Figures 2.15 and 2.16).

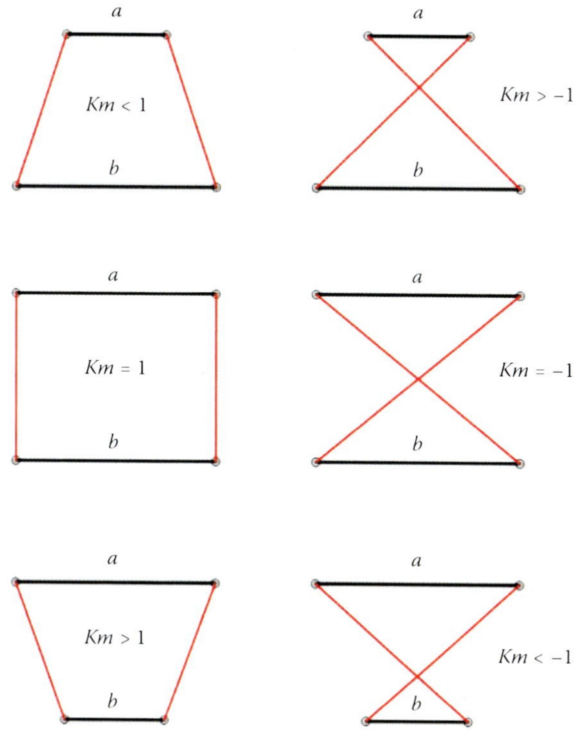

FIGURE 2.14 Compensator suspensions schemes.

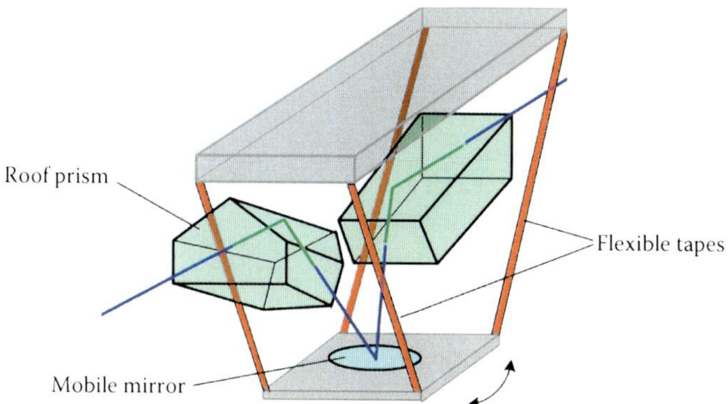

FIGURE 2.15 Compensator with an Abbe prism.

5. Using flexible tapes of different elasticity for a bracket suspension and changing the flexible tapes' length. The compensator sensitivity increases depending on the flexible tapes' elasticity. The tapes' elasticity depends on their physical properties and also the force applied to them. The force depends on the compensator movable part weight. In case the tapes are

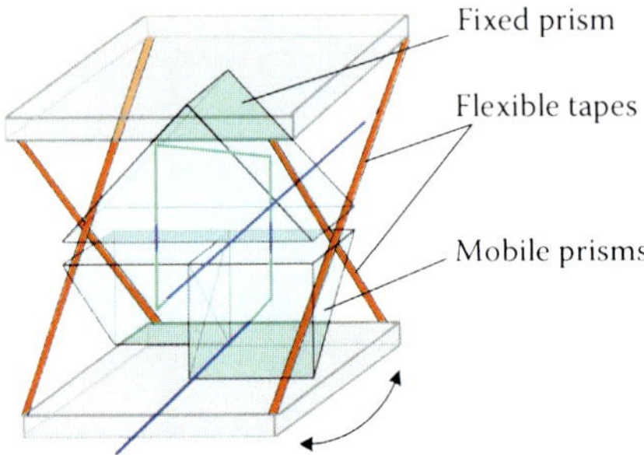

FIGURE 2.16 Compensator with Porro prism.

more elastic, the sensitivity increases. Also the compensator sensitivity increases depending on the tapes' length. We shall use Kp as the compensation factor that depends on the tapes' length and elasticity.

The total formula for the level compensator that fulfills an optical image inversion is

$$K = -Km\,Ko\,Kp \tag{2.2}$$

The minus sign (–) means the optical system is inverting.

Now we will consider the arrangement of a typical automatic level (Figure 2.17). The telescope is almost the same as a modern dumpy level, except the inverting

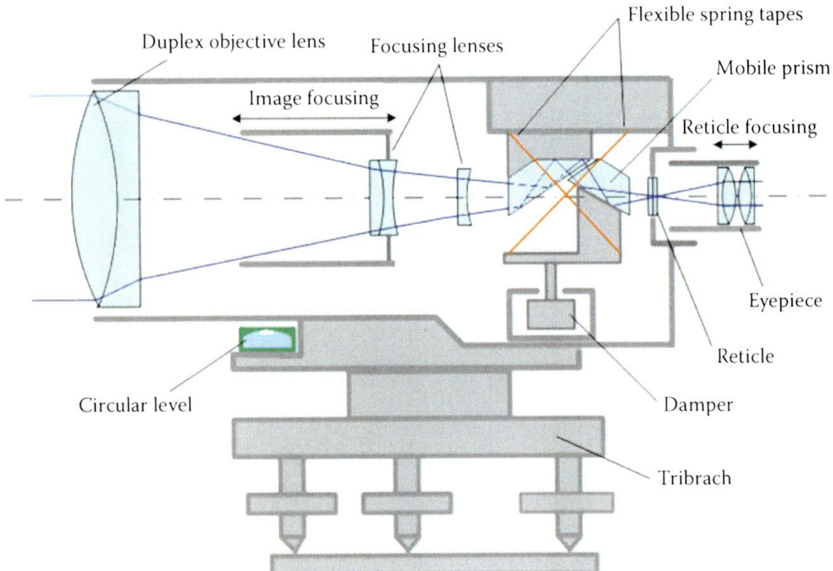

FIGURE 2.17 Automatic optical level.

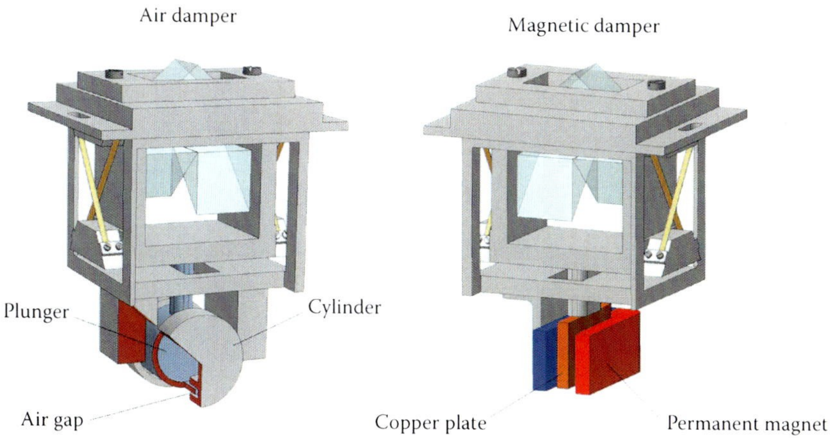

FIGURE 2.18 Compensator's dampers.

optical system is replaced with the compensator. The pendulum of the compensator is equipped with a damper. Also the adjusting screws appear to move the reticle in a vertical direction. The other parts have become less complicated than those similarly found on a dumpy level. Thus, due to the absence of a tubular level leveling mechanism, the base structure has become simpler. The tribrach with its foot screws remained the same. The circular level is also in the same place.

Air or magnetic dampers are applied in modern automatic levels for damping from free pendulum fluctuations (Figure 2.18). The principle behind an air damper relies on the air friction created in the small gap between a cylinder and the plunger that is set on the compensator pendulum. Such a small gap demands exact manufacturing and proper adjusting. The magnetic damper operates in a different way. A copper plate is set on the compensator pendulum. There is an interaction between the magnetic fields from a permanent magnet with the magnetic fields of eddy currents. These currents (also called Foucault currents) occur in the copper plate that moves in the permanent magnetic field. The interaction creates an electromagnetic force that is directed inversely to the copper plate's movement damping the free pendulum's fluctuations.

Now we look into the features and use of automatic levels. Let us suppose one level has been slightly dropped. External damages are not present. Is it still possible to use the level? For dumpy level users the answer is evident: check whether collimation and a tubular level axes are parallel. If everything is all right, it is still possible to use the dumpy level. Nearly everything is more difficult regarding automatic levels. Here it is necessary to analyze the compensator's condition.

It is better to do research by the means of a special collimator whose optical axis is set into a horizontal position. The collimator is a long focus telescope (usually from 400 to 600 mm) and adjusted to infinity. The collimator axis is set horizontally by the means of other precise instruments. The collimator reticle has the scale divisions graduated at 20″ or 30″ intervals. A collimator for checking surveyor's instruments usually has a regulated height adaptor for installation of these devices.

If we have no collimator it is necessary to make a test stand. In that case, two identical high-quality rulers graduated at 1 mm intervals are required. These rulers are fixed on opposite walls at the same height. It is carried out with the aid of a carefully calibrated surveyor's level set into the center of the room. The distances from the rulers to the instrument should be from 10 to 15 m and equal each other. The stand is ready.

We put a tripod at a short distance (about 1 m) from one of the rulers and fix the tested level so that one of the foot screws is directed to the other distant ruler. We sight at the near ruler and read out the value. Then, we look at the distant ruler and read out the value again and then compare these two values. And if they are different, the distant ruler result should be corrected by the means of the reticle adjusting screws so that this value is equal to the nearer ruler value. Thereby the level has been checked and corrected only at one position of the compensator when the bubble of the circular level is in the middle.

Next we test the compensator operation at different inclinations. First, we test the compensator while inclining the telescope along the collimation axis. So the values on both rulers are equal when the circular bubble is in the center. In the left side of Figure 2.19 these values are 43.5 mm for instance. Let us take it as referential value. If the compensator operates correctly at longitudinal inclination of the level, the reticle horizontal line should not move from the referential value.

We incline the telescope toward the distant ruler's direction by the means of the foot screw. If the telescope inclination is fulfilled upward (the middle of Figure 2.19) and the reticle horizontal line is above the referential value, then there is an insufficient compensator sensitivity (undercompensation). If the horizontal line is below the referential value (the right side of the figure) that means there is an extra sensitivity of the compensator (overcompensation).

There are two ways to influence the sensitivity of the compensator. The first way is moving the compensator along the telescope collimation axis (the left side of Figure 2.20). According to Equation 2.1, the compensator sensitivity increases if we move the compensator frame toward the objective lens. In case of overcompensation, we should move the compensator back nearer to the eyepiece. The second way is changing the compensators center of gravity. There is an adjusting weight for

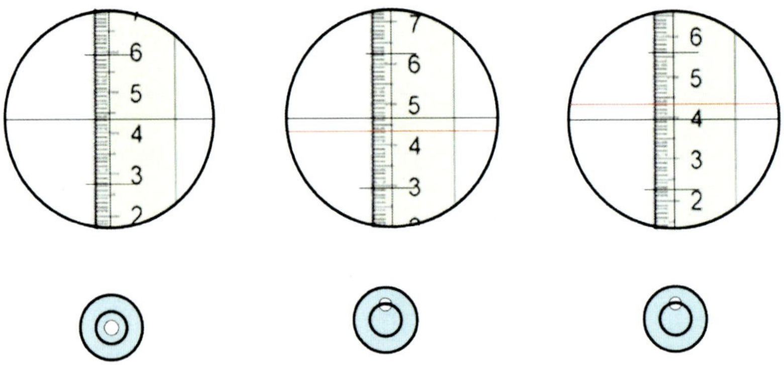

FIGURE 2.19 Checking of longitudinal compensator stability.

FIGURE 2.20 Adjusting of compensator position.

compensator sensitivity adjustment. If we move the adjusting weight up, the compensator sensitivity increases. After we move the compensator or the adjusting weight, we should move the circular level bubble back to the middle and adjust the distant ruler value to a referential one by means of the reticle adjusting screws.

Further, we repeat the process several times until the reticle horizontal line is on the referential value at all longitudinal compensator inclinations. We should complete the adjustments while inclining the telescope to different directions within the compensator operation range. If the compensator operates correctly at one inclination but at the opposite inclination there is undercompensation (or overcompensation), then the flexible tapes suffer from deformation. It is necessary to bring this level to a workshop for tape replacement.

After we have adjusted the compensator sensitivity we should verify the compensator's reaction to lateral inclinations. The preparation to this test is similar to the previous case. However, the tripod must be turned over at 90° relative to the prior position. We incline the instrument to the transverse direction by the means of the foot screw. We watch the direction and the value of the instrument inclination with the aid of the circular level. Of course the reticle horizontal line is also inclined, which is why we should read out the ruler closer to the scale center.

When the compensator is set parallel to the collimation axis, the center of the reticle should not be shifted from the referential value at lateral inclinations. If the shift occurs (Figure 2.21), then it is necessary to rotate the compensator around its vertical axis (right side of Figure 2.20). It is important to remember the following: if we incline the instrument to the left and the reticle horizontal line rises above the referential value (left side of Figure 2.21), it is then necessary to turn the compensator clockwise around its vertical axis. If the reticle horizontal line is below the referential value (see the right side of Figure 2.21), then we should turn the compensator to the opposite direction.

After adjusting of the compensators position we must sight both at the near and the distant rulers and read out the values. If the values are different we should correct the distant ruler readings by the means of the reticle adjusting screws.

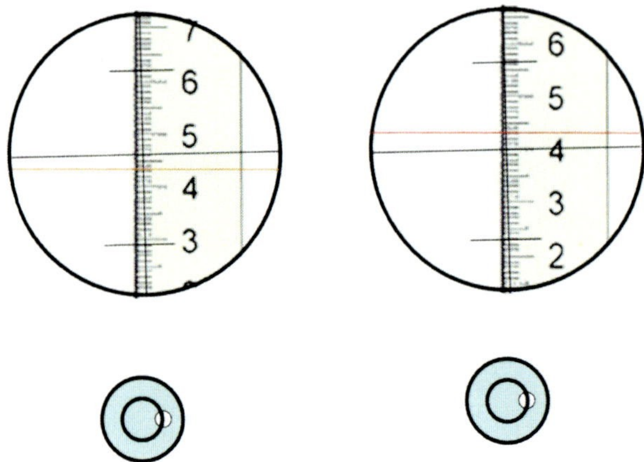

FIGURE 2.21 Checking of lateral compensator stability.

TABLE 2.2
Up-to-Date Automatic Levels

Model	Accuracy (mm/km)	Magnification (n×)	Compensator Accuracy/ Operating Range (n''/n′)	Minimal Focusing Range (m)	Manufacturer
NAK2	0.3–0.7[a]	32–40	0.3/±15	1.6	Leica Geosystems
B20	0.5–0.7[a]	32	0.3/±15	0.2	Sokkia
AS-2S	0.4–0.8[a]	34	0.3/±12	1	Nikon
AE-7C	0.45–1.0[a]	30	0.35/±16	0.3	Nikon
NA730	1.2	30	0.3/±15	0.7	Leica Geosystems
B30	1.5	28	0.5/±15	0.3	Sokkia
NA728	1.5	28	0.3/±15	0.5	Leica Geosystems
AP-8	1.5	28	0.5/±16	0.75	Nikon
NA724	2	24	0.5/±15	0.5	Leica Geosystems
B40	2	24	0.5/±15	0.3	Sokkia
AS-2S	2	24	0.5/±16	0.75	Nikon
NA720	2.5	20	0.5/±15	0.5	Leica Geosystems
AX-2S	2.5	20	0.5/±16	0.75	Nikon

[a] With micrometer.

Nowadays automatic levels have become the most widely spread type of surveying levels. It is difficult to list all of the modern levels, but we must mention the instruments that have become the prototypes of the subsequent forms developed by different brand manufacturers (Table 2.2).

2.1.2.2 Autofocusing Levels

Within the last 10 years, automatic levels with autofocusing (AF) have entered the market. In Figure 2.22, we can see a typical automatic level, modified with new AF parts.

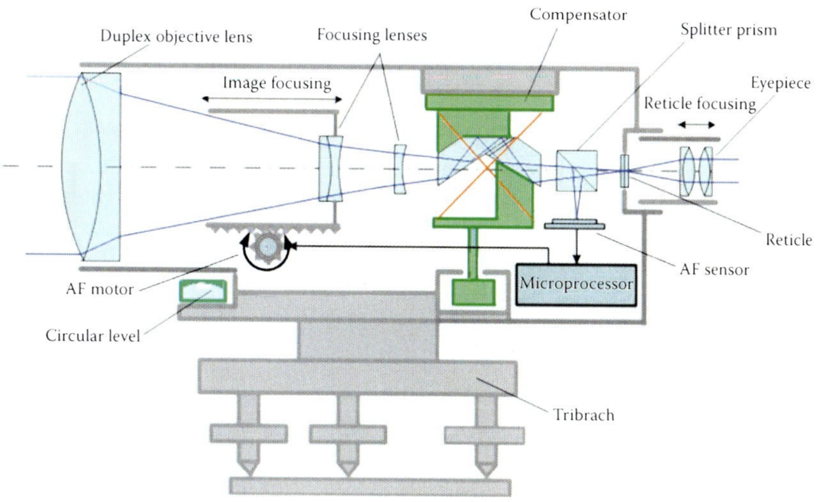

FIGURE 2.22 Autofocusing level.

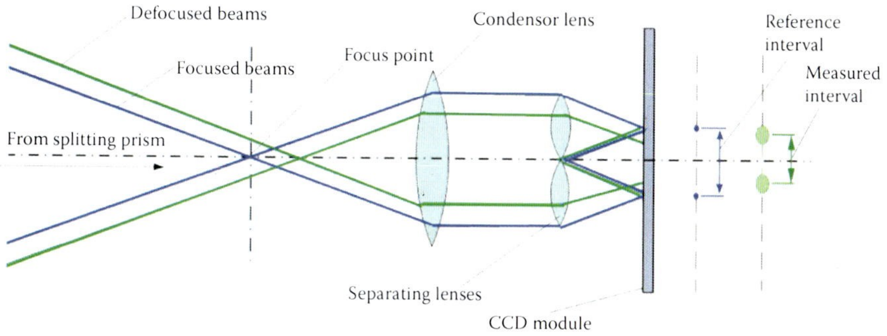

FIGURE 2.23 Autofocusing operation principle.

In the optical scheme of an autofocusing level there is a splitting prism located between the compensator and the reticle. Part of the optical beam is reflected by its internal semitransparent mirror surface and directed to the AF sensor. The AF sensor module consists of a condenser lens, a couple of separating lenses, and a photosensitive charge-coupled device (CCD) module (a CCD matrix or a CCD line).

The condenser lens transforms the converging optical beams to the parallel ones. These beams are directed to the pair of separating lenses. There is a CCD photodetector within the focus range of these lenses. If the image of the rod is focused (Figure 2.23), then the pair of images on the CCD line has accurate contours and maximal contrast. If the image is defocused, then the contrast decreases, and the interval between these images on a CCD module is affected. The signal moves from the CCD module to the microprocessor, which processes it and finds the interval between the images. If this interval does not correspond to the reference interval, the microprocessor sends a command to the AF motor driver. The AF motor moves the focusing lens until the measured interval equals the referential one.

Autofocusing levels can also be used in a manual focusing mode. They have focusing handles, as do the manual focus automatic levels.

It is necessary to verify the accuracy of coincidence between the AF unit optical axis and the vertical line of the reticle, and also confirm the coordination of the automatic and manual focusing. Both parameters can be checked simultaneously. We begin by placing a leveling rod at distance of about 20 m from the autofocusing level. The rod is set in front of a bright wall. We determine the intervals to the left and to the right from the rod's center where the autofocus still operates. The difference between these intervals should be less than 50%. It is possible to reduce this difference by moving the reticle horizontally, if the level design allows that.

The automatic and manual focus coordination is easy to check. Once the autofocus is used we try to acquire a more focused image by means of the manual focusing knob. If while rotating this knob in either direction the image's focus worsens, then the autofocus is confirmed as working correctly. Otherwise, it is then necessary to shift the sensor AF, but that should be done at a service center.

Some types of autofocusing levels are shown in Table 2.3.

2.1.2.3 Digital Automatic Levels

The basis for a digital level is an automatic level. However, a digital level is added with an automated rod value reading system. This system consists of a splitting prism, a CCD line, and a microprocessor. The reading system uses special bar-coding rods. At the present moment there are four different digital rod standards in which different digital data coding are applied (Figure 2.24). Digital bar coding is similar to a usual rod division into decimeter parts and "coding" them by the means of Arabian figures.

Digital leveling strategy is similar to that of analog levels. It consists of determining measured results in two steps. The first step is finding an approximate value.

TABLE 2.3
Modern Autofocusing Levels

Model	Accuracy (mm/km)	Magnification ($n\times$)	Compensator Accuracy/ Operating Range (n''/n')	Minimal Focusing Range (m)	Manufacturer
SDL1X	0.2–1.0[a]	32	0.3/±12	1.6	Sokkia
AFL-321	0.4–0.8[b]	32	0.3/±12	0.6	Pentax
AFL-281	1.5	28	0.5/±12	0.6	Pentax
AFL-241	2	24	0.5/±12	0.6	Pentax

[a] Special invar rod.
[b] With micrometer.

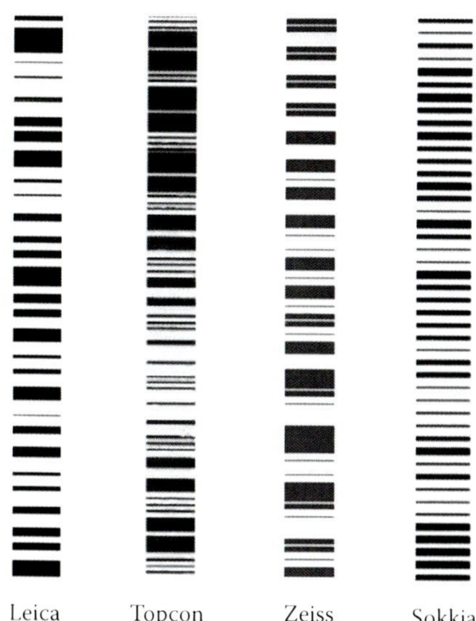

FIGURE 2.24 Rods for digital levels.

A usual rod is coded with Arabic figures and centimetric marks. In a digital rod the approximate value is coded using a combination of white and black strips of various widths. The second step consists in finding an exact value. In a usual level we would assess a millimetric distance from the horizontal line of the reticle to the nearest lower edge of the centimetric mark. In a digital level the CCD line pixels are calculated from the "zero" pixel to the central pixel of the white stripe that is situated below. The final measuring result is the microprocessor integration of these steps.

Let us learn the scheme of digital level (Figure 2.25). The digital level consists of similar components as an automatic level. If we switch off the electronic part, the digital level can still work as an automatic level with usual leveling rods. The optical scheme of a digital level is added with a splitting prism as well as an autofocusing level. The photosensitive CCD line is located in the telescope's focal plane directly behind the reflecting facet of the splitter prism.

In Figure 2.26, we can see the field of view of the CCD line where the digital rod image is projected. An electric analog signal at the output of the CCD line goes to the microprocessor for processing. The microprocessor has the analog/digital (A/D) converter, which normalizes the analog signal and transforms it into the digital sequence. Furthermore, the microprocessor finds the central position of each impulse. The impulse centers correspond to the centers of the white strips of the barcode rod. The interval (P) corresponds to the grid step of the leveling rod. Also, the microprocessor calculates the distance (f) from the impulse center to the zero pixel of the CCD line. The position of the zero pixel is set by the microprocessor program so that it coincides with the reticle horizontal line. The next step is the microprocessor's calculation of an exact part of the readout by dividing f into P. It is similar to our reading of millimeters from the usual leveling rod as a part of the centimeter mark.

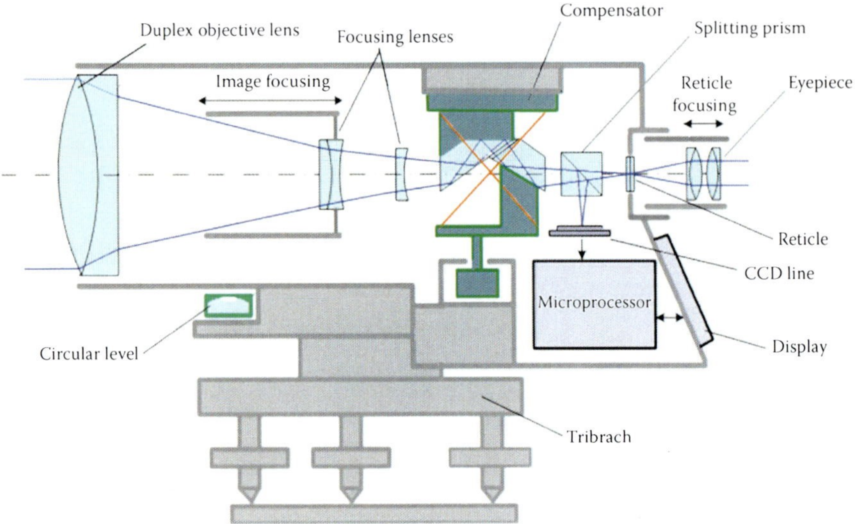

FIGURE 2.25 Digital level.

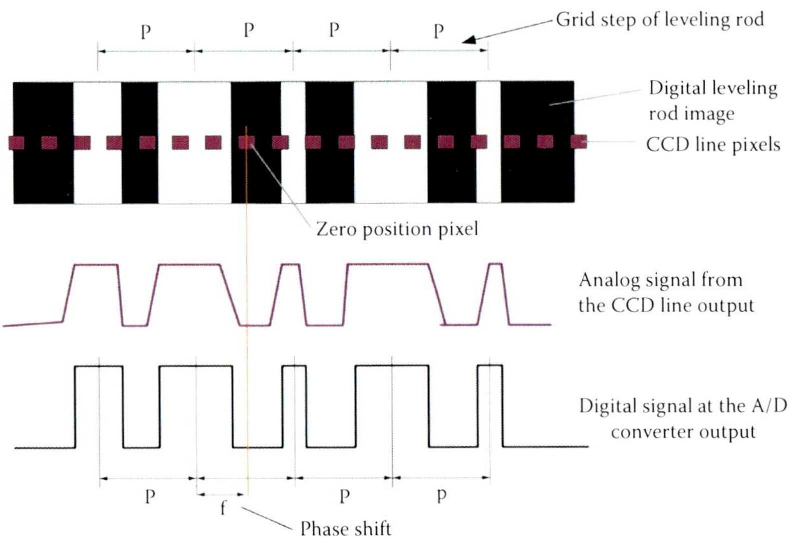

FIGURE 2.26 Readout principle of a digital rod.

There are two ways to find the approximate readout value of the digital rod. In the first case, the digital rod is divided into parts that are coded as nonrepeating combinations of the black and white strips of the barcode. There is a record of these barcoding combinations in the memory of the microprocessor represented in a table. The result is acquired by comparing the current value with the tabular one. The

TABLE 2.4
Up-to-Date Digital Levels

Model	Accuracy (mm/km)	Magnification ($n\times$)	Compensator Accuracy/ Operating Range (n''/n')	Minimal Focusing Range (m)	Manufacturer
SDL1X	0.2–1.0[a]	32	0.3/±12	1.6	Sokkia
DiNi 03	0.3–1.0[a]	32	0.2/±15	0.2	Trimble
DNA03	0.3–1.0[a]	24	0.3/±10	0.6	Leica Geosystems
SDL30	0.4–1.0[a]	32	—/±15	1.5	Sokkia
DL-502	0.6–1.0[a]	32	—/±15	1.5	Topcon
SDL50	0.6–1.0[a]	28	—/±15	0.7	Sokkia
DiNi 07	0.7–1.3[a]	26	0.5/±15	0.3	Trimble
Sprinter 250M	0.7–1.0[a]	24	—/±10	2.0[b]	Leica Geosystems
DL-503	0.8–1.5[a]	28	—/±15	1.5	Topcon
DNA10	0.9–1.5[a]	24	0.8/±10	0.6	Leica Geosystems
DL201	1	32	—/±12	1.5[b]	South
Sprinter 150M	1.5	24	—/±15	2.0[b]	Leica Geosystems
DL-202	1.5	32	—/±16	1.5[b]	South
EL28	1.5	28	0.4/±15	0.8	Foif
Sprinter 50	2	24	—/±10	2.0[b]	Leica Geosystems

[a] With invar rod.
[b] Digital measure mode.

second way does not demand the table in the microprocessor memory. There are two groups of strips. Each group has the strips arranged sequentially with regular intervals, and the strips' width smoothly variates like a sine curve. The width variation occurs differently. They have periods of 600 and 570 mm, with a phase shift of $\pi/2$ at the beginning of the rod. Such a combination enables one to determine nonrecurrent approximate readouts at the rod length of 4 m. Further, the microprocessor unites the exact and the approximate readouts while sending the results to the display.

Some types of digital levels are shown in Table 2.4.

The compensator of a digital level is tested in the same way as an optical automatic level compensator. The digital portion is switched off in this case.

There are two features for assessing the digital part. One of them is checking a symmetric operating range relative to the reticle vertical line. It is similar to an autofocusing level verification as described earlier. The other feature consists of gauging the coincidence between a horizontal line and the zero pixel of the CCD line. It is necessary to check the coincidence of readouts on the digital and analog leveling rod sides. In case of noncoincidence we need to correct the zero pixel position. There is a special program in the digital level menu to complete this step.

2.2 LASER LEVELS

2.2.1 Principle of Laser Level Operation

A laser level is a device that creates either a visible laser line or plane, perpendicular to the plumb line. A laser level consists of the laser module and a leveling mechanism. The red laser module (Figure 2.27) has a laser light source (a red laser diode) and a condenser lens. The laser diode emits a divergence beam that becomes parallel when it passes through the condenser lens. The laser-emitting crystal must be accurately sets within the condenser lens's focus. The laser crystal is placed on a heat sink next to the controlling photodiode. The principle of laser crystal operation is covered in Chapter 4. Now we should note that laser light power strongly depends on the crystal temperature. In order to minimize this dependence a laser driver with feedback control is used. The laser crystal emits light in opposite ways. Luminous power from the opposite side of the crystal enters the integrated photodiode. The photodiode signal depends on the laser's luminous power. The signal enters the laser driver's input. The amplified signal controls the power regulator so that the laser's luminous power is stable. The luminous power from the laser crystal's front side looks like a prolate ellipse. The condenser lens's diameter overlaps only the central zone of this ellipse where a majority of the luminous power is concentrated. The laser beam from the condenser output can be directly applied or be transformed into a laser plane by the means of the cylindrical lens (Figure 2.27).

Sometimes green lasers are used in laser levels. The purpose is that the relative spectral sensitivity of an eye considering green light is approximately four times higher than that of red. Green laser diodes have already been developed. However, they are still very expensive and are not currently applied in laser levels. Green laser modules where green light is transformed from an infrared source are still widely

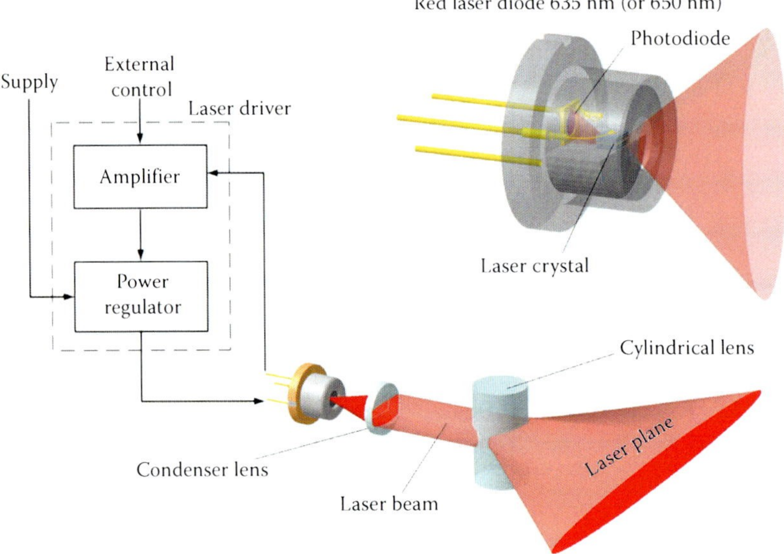

FIGURE 2.27 Red laser module provided with a cylindrical lens.

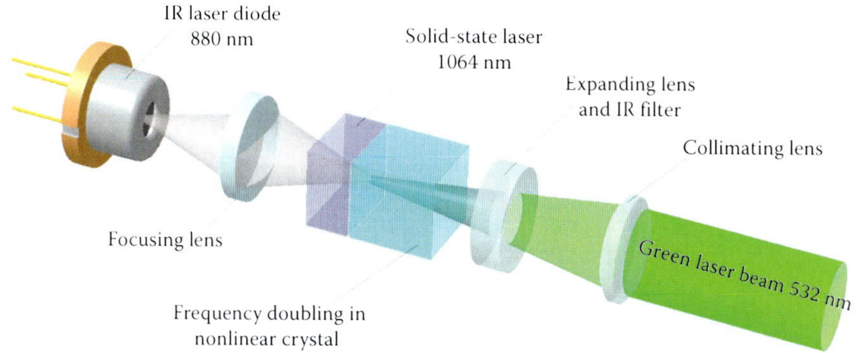

FIGURE 2.28 Green laser module.

used (Figure 2.28). Such laser modules are easy to produce, and as they cost a bit more than red laser ones, they are ten times cheaper than green laser diodes. The light source for a green module is an 880 nm infrared laser. The light enters the solid-state laser, which generates light of 1064 nm wavelength. Then that light enters a nonlinear crystal that redoubles the light wavelength up to 532 nm. A mix of green and infrared lights occurs at the output of a nonlinear crystal, which then enters the expanding lens, acting as the infrared filter as well. The collimating lens transforms dispersed light into a parallel green laser beam.

The efficiency of green laser modules is lower than red ones because of the complicated energy transformation. The optimal operating temperature range for green modules is also narrower than their red counterparts. In the case a laser level is used with a laser plane sensor, application of a red laser suggests the sensor's sensitivity to red light is higher than green.

It is possible to transform a laser beam into a laser plane by the means of a mirror cone that creates a laser plane within 360° (Figure 2.29). A cylindrical lens only allows a quality plane within 90°, but a laser trace from the laser plane is brighter

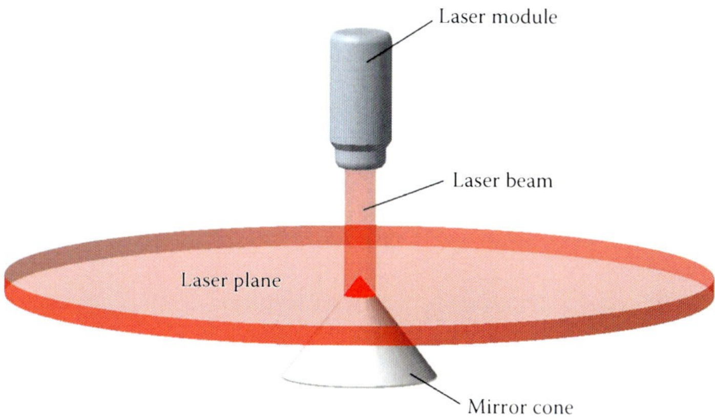

FIGURE 2.29 Laser module provided with a mirror cone.

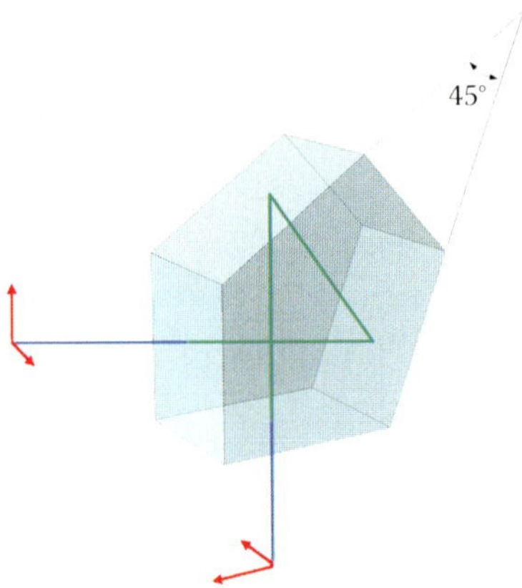

FIGURE 2.30 Pentaprism.

than the trace created by means of a mirror cone. That is why a cylindrical lens is used more often than a mirror cone to transform the beam.

The mechanical conversion of a laser beam into a plane uses a rotating optical element. Rotating elements that can be used include a mirror inclined at 45° to the beam axis, an elementary right-angled prism, a roof prism, or a pentaprism (Figure 2.30). The pentaprism is often applied as the main element of a mechanical converter. The pentaprism stabilizes the beam deflection in a case of the roller bearings whipping. The leveling mechanism may be manual or automatic. The manual mechanism consists of a platform that has two perpendicularly located tubular levels (Figure 2.31). The platform has a holder with laser modules that transform laser lines into planes. The elementary solution has two modules: both horizontal and vertical. The most complicated units may have as many as four vertical and three horizontal modules, and another with a linear laser plumb. The platform may be inclined in two planes by the means of two leveling screws or the three foot screws of the tribrach.

There are two types of automatic leveling mechanisms: leveling motor-driven mechanisms and pendulum self-leveling mechanisms. The first type of mechanism is based on "hand" leveling principles where the leveling screws are rotated by the motors instead of hands (Figure 2.32).

Laser levels may have motors with a reducer or step-motors without any reducer. These motors are controlled by the motor driver, which receives signals from the inclination sensors. There are two types of sensors. Either mono-axial or dual-axial sensors are used in laser levels. Combinations of two tubular levels are set perpendicularly to each other or one circular level is applied within an auto-leveling mechanism. The sensor consists of a sensitive cell and an electronic unit. The sensitive cell is a tubular or circular level filled with electrolyte (Figure 2.33). There are electrodes on the inner or the external surface of the level vial. The electrodes are metallized segments. Depending on the electrodes arrangement, the sensors act as capacitance sensors or impedance sensors. The capacitance sensor has external metallized segments, which together with electrolytes form variable capacitors (Figure 2.34). Their capacity depends on the position of the level bubble. When the capacitor is cut in

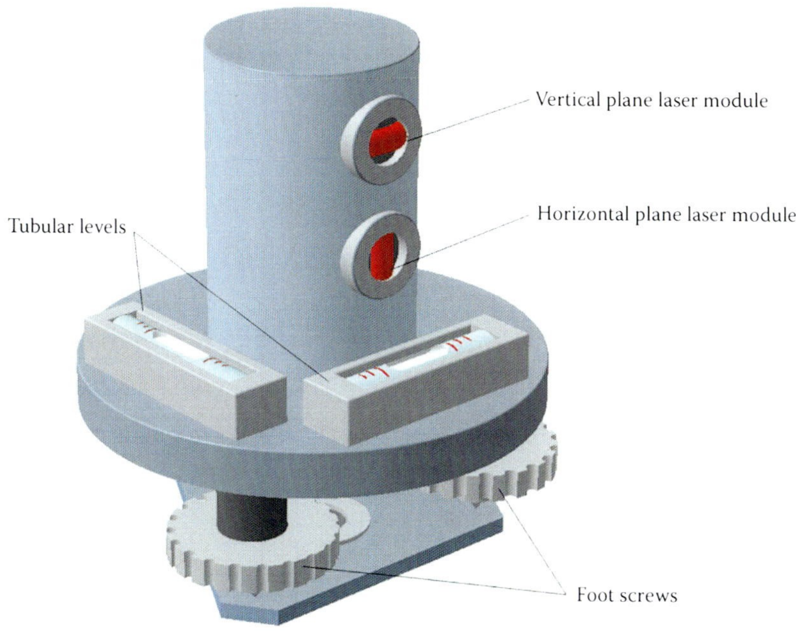

FIGURE 2.31 Nonautomatic laser level.

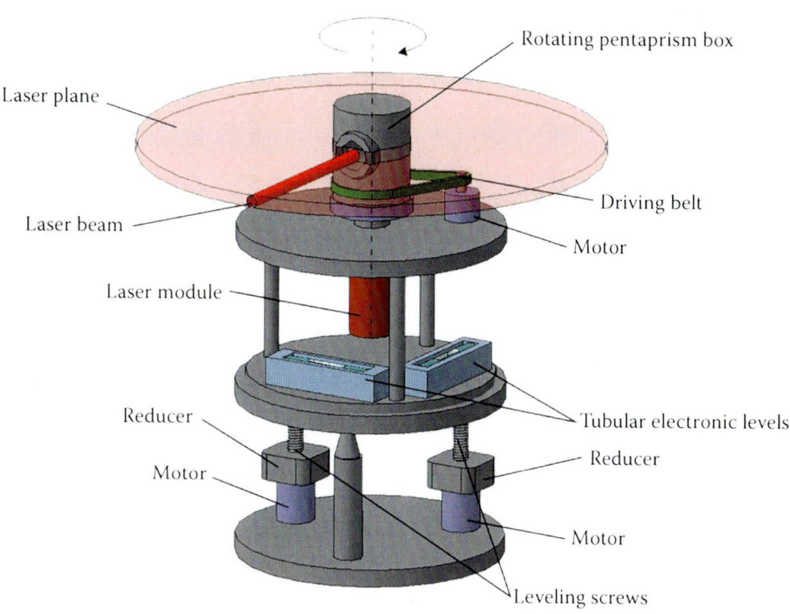

FIGURE 2.32 Rotary automatic laser level.

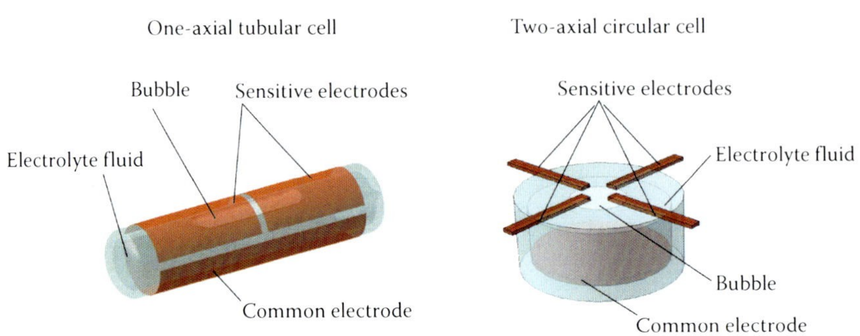

FIGURE 2.33 Sensitive cells of an electronic level.

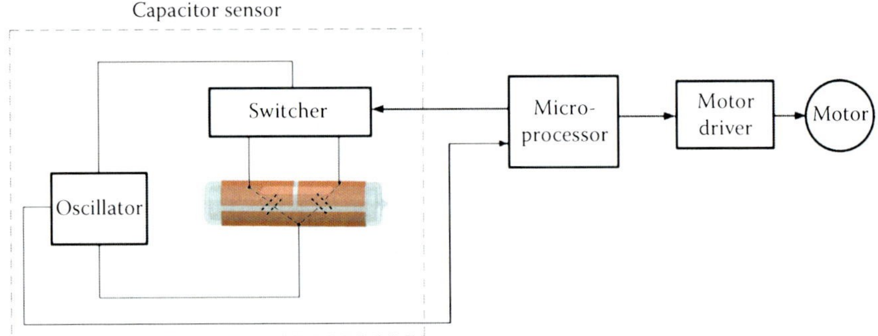

FIGURE 2.34 Auto leveling principle with the capacitive tubular cell.

a frequency-dependent circuit of an oscillator its frequency depends on the bubble position. The microprocessor controls the switcher alternatively connecting every capacitor to the oscillator and measures the oscillator frequency at each step.

When the level bubble is not in the center of the vial, the oscillator frequency is different at both positions of the switcher. The microprocessor generates a signal that is sent to the motor driver that instructs the leveling screw to rotate in the required direction. It also inclines the platform with the sensors to an opposite direction until the difference of frequencies disappears.

The electrodes of the impedance sensor are located in the vial. The resistance and capacity occurring between the electrodes depend on the bubble position. An example of using of the impedance cell is shown in Figure 2.35. The antiphase pulse signals enter the upper electrodes. A resulting signal occurs in the common electrode. Its phase and amplitude depend on the position of the bubble of the level. The preamplified signal enters the synchronous detector, which consists of the demodulator and a filter. In the synchronous detector this signal is compared to a basic signal from the oscillator. The direct voltage signal appears at the filter output. Polarity and amplitude of this signal depends on the sensor inclination. The motor rotates the leveling screw to minimize the signal at the filter output.

Step motors are used in the top-end models of rotational levels for pentaprism rotation. The rotary mechanism provided with the step motors allows operation in three

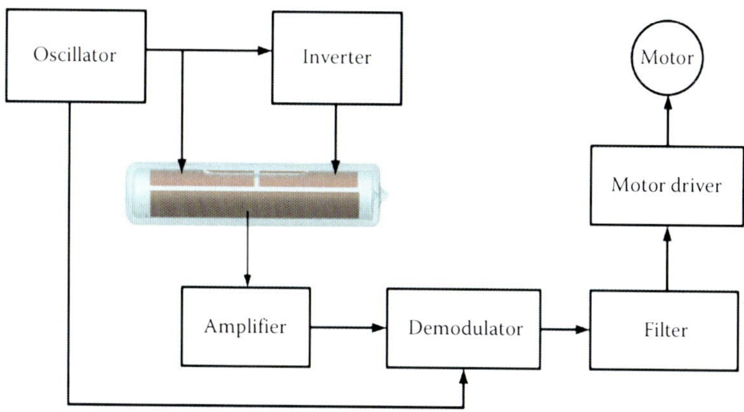

FIGURE 2.35 Auto leveling principle with the impedance tubular cell.

modes: horizontal laser line mode, laser sector mode, and full laser plane mode. The full laser plane mode is used in places with weak background lighting. Outside it is preferred to use narrow laser sector mode simultaneously with a laser plane sensor (Figure 2.36).

A laser plane sensor can be used in combination with an ordinary leveling rod. The sensor has two vertically installed photodetectors that are connected to high-sensitivity amplifiers. Signals from these amplifiers enter the logic processor, which has a simple display of its output. It indicates presence of laser light at the sensors. If these signals from the photodetectors are equal, it indicates the presence of the laser plane between the photodetectors. The event is indicated on the display and noted by a sonic call. There is a monochromatic light filter in the front of photodetectors to diminish influence of background lighting.

There are two kinds of pendulum dual-axial self-leveling mechanisms: direct or inverse pendulum. The direct pendulum solution is often used with a cardan

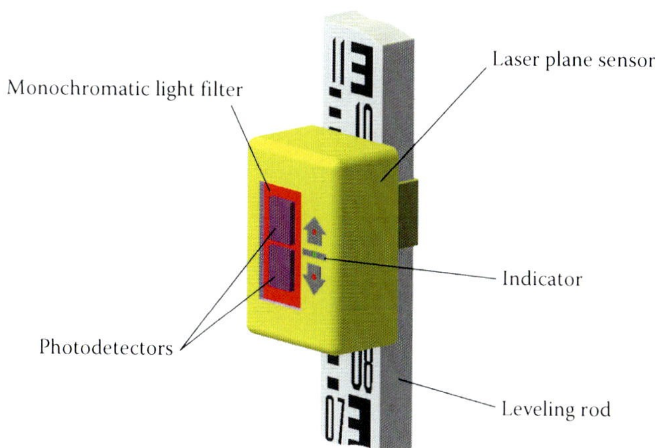

FIGURE 2.36 Laser plane sensor on the leveling rod.

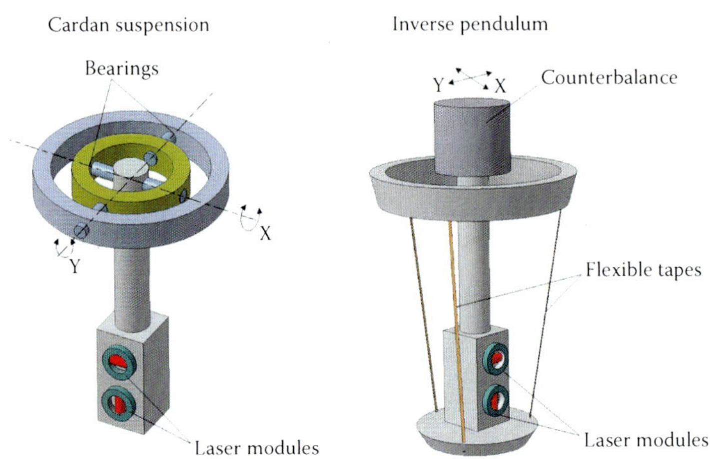

FIGURE 2.37 Pendulum self-leveling mechanisms.

suspension, and the inverse pendulum solution uses a flexible suspension (Figure 2.37). The pendulum is fitted with laser modules that use optical converters "line–plane." In the figure we can see a minimal set of laser modules: the vertical and the horizontal ones. The manufacturer can install several additional modules: up to four vertical and three horizontal.

A magnetic dumper is set in the bottom part of the pendulum. The pendulum suspensions are very delicate mechanisms. Therefore, in these types of laser levels there is a lever that fixes the pendulum when the power is switched off. Some types of laser levels are shown in Table 2.5.

TABLE 2.5
Up-to-Date Laser Levels

Model	Accuracy (±mm/m)	Range (Sensor) (m)	Leveling Mechanism	Laser Color	Beam Sweep	Manufacturer
Rugby 100RL	1.5/30	750	Motors	Red	Rotary	Leica Geosystems
UL633	1.5/30	800	Motors	Red	Rotary	Spectra Precision
HV401	1.5/30	400	Motors	Red	Rotary	Spectra Precision
HV301G	2.2/30	250	Motors	Green	Rotary	Spectra Precision
Roteo 20HV	3.0/30	150	Motors	Red	Rotary	Leica Geosystems
Roteo 35G	3.0/30	150	Motors	Green	Rotary	Leica Geosystems
LL20	1.5/10	80	Pendulum	Red	Mirror cone	CST/Berger
GLL3-80	2.0/10	80	Pendulum	Red	Mirror cone	Bosch
Lino L2+	2.0/10	30	Pendulum	Red	Mirror cone	Leica Geosystems
Lino L2	3.0/10	30	Pendulum	Red	Cylindrical lens	Leica Geosystems
GLL2-15	3.0/10	15	Pendulum	Red	Cylindrical lens	Bosch
EL-503	3.0/10	20	Manual	Red	Rotary	Geo Fennel

2.2.2 Checking and Adjusting of Laser Levels

In the factory, laser level adjustment is accomplished by the means of using collimators linked with video cameras. We are describing easy and effective ways for checking and adjustment of a laser level in a small servicing center or user's enterprise. Therefore, we need a close to square room with a stable floor. The optimal distance between the walls is 10 m. It is advisable to have a faint light in the room or some curtains at the windows. We then set up an accurate automatic optical level into the center of the room, and then stick some marks at the same height on every wall of the room. Then we set a theodolite on the tripod instead of the optical level and lay out a vertical plane by the means of two marks on one of the walls. We should complete lay out of the vertical plane twice and at two positions of the theodolite. We also need to have a tripod with an elevating platform. Although not necessary, the use of simple theatrical binoculars will aid the process. Now everything is ready!

We shall start working with the levels where the cylindrical lenses are used in laser modules. Such a laser module creates a proper laser plane on the condition that the cylindrical lens is set strongly perpendicular to the axis of the laser beam. This state is appropriated by the manufacturer of the laser module, but the correlation would be affected if the instrument is dropped or after unskilled repair. First, we should seek to correct laser plane curvature occurrences. They are the result of the cylindrical lens being incorrectly positioned. Now we put the instrument on the elevating tripod so that it is placed next to the wall opposite of the wall with our vertical marks (Figure 2.38). Now, we switch on the horizontal and vertical plane of the laser level pressing the buttons on the controls and point it toward the center of the opposite wall. The lifting screw is used to aim the height. Then we start adjusting the first horizontal plane, if there are several horizontal modules. The horizontal laser line can be projected onto the wall with marks in one of three forms (Figure 2.39).

If a curvature of the laser plane appears we should adjust the cylindrical lens position in the horizontal module. It is carried out by means of a pair of adjusting screws

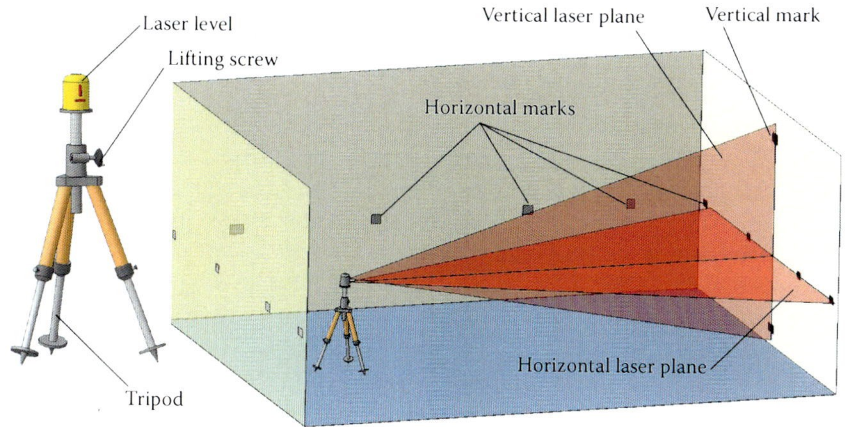

FIGURE 2.38 Checking and adjusting of a laser level.

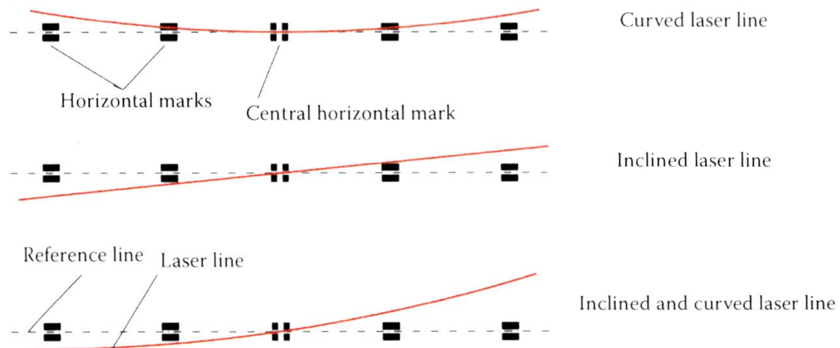

FIGURE 2.39 Laser plane checking.

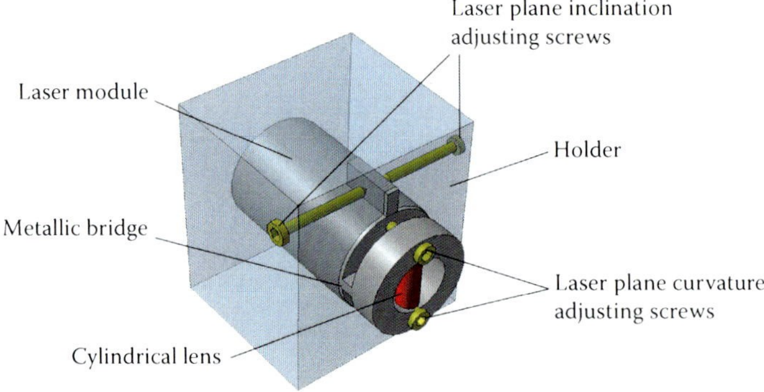

FIGURE 2.40 Laser module adjusting screws.

(Figure 2.40). Carefully we loosen one adjusting screw while tightening the other. As a result, a metallic bridge is bent and the cylindrical lens is inclined. It is advisable to watch the laser line on the distant wall using a theatrical binocular.

Having corrected the laser line curvature we move on to correct the lateral inclination of the line. Let us superpose at the height of the laser line with the central horizontal mark on the opposite wall. The horizontal laser line must coincide with the row of horizontal marks on this wall. A noncoincidence indicated a lateral inclination of the laser plane. The reason for an inclination may have occurred during the laser module inclination as incorrect operation of the self-leveling unit of the laser level. In this case, we need to carefully analyze the cause. Usually a laser level has several laser modules, and each of them generates its own laser plane. In order to find out the source of errors we will need to look at the other laser modules. If they are accurate, then we should correct the first horizontal laser module's inclination. This is done by means of a pair of adjusting screws (see Figure 2.40). By screwing and unscrewing them we can lightly rotate the laser module about the holder.

If most of the planes are inclined, we should correct the inclination by adjusting the self-leveling mechanism, either the mechanism provided with a cardan suspension or having an inverse pendulum; both are adjusted with the aid of adjustment weights.

If the instruments use an automatic moto-leveling platform, the laser planes inclinations are adjusted by inclining the capacitor's sensors (see Figure 2.33). The adjustment can also be fulfilled using an electronic approach. There are special trimmers in the electronic board to do this. The most advanced and expensive laser levels are adjusted through software.

Having corrected inclination and curvature of the first horizontal laser plane, we then start correcting its longitudinal inclination. We turn the instrument at 180° and superpose the laser line with the horizontal marks on the nearest wall. Then we move the instrument to direct it to the opposite wall. If the horizontal lines coincide with the marks, it means the self-leveling mechanism is correctly adjusted. If they do not coincide there is longitudinal inclination of the horizontal plane. It can be corrected by changing the position and weight of special adjusting weights. Now we assess the laser plane's inclination, which is created by other laser modules, if they are set in the instrument.

Verification and adjustment of rotary laser levels is very easy using our room. We place the instrument on the tripod, which is placed in the center of the room. Then we switch on the instrument and superpose the laser plane at the height marks on one of the walls. If the plane coincides with the marks on the other walls, then the self-leveling mechanism operates correctly. If not, that means the laser plane is inclined (Figure 2.41).

Detected inclinations of the laser plane can be corrected by adjusting the leveling mechanism as described earlier by the means of trimmers, software, or electronic sensors inclinations.

After adjusting, an automatic laser level is tested through its whole operating range of inclinations. We can set these inclinations via the foot screws. Some elementary models of laser levels may not have foot screws. In this case we can use a theodolite tribrach and an adaptor. Looking into the instrument's manual we can see its operating range. Before the start of testing we should put the instrument into the

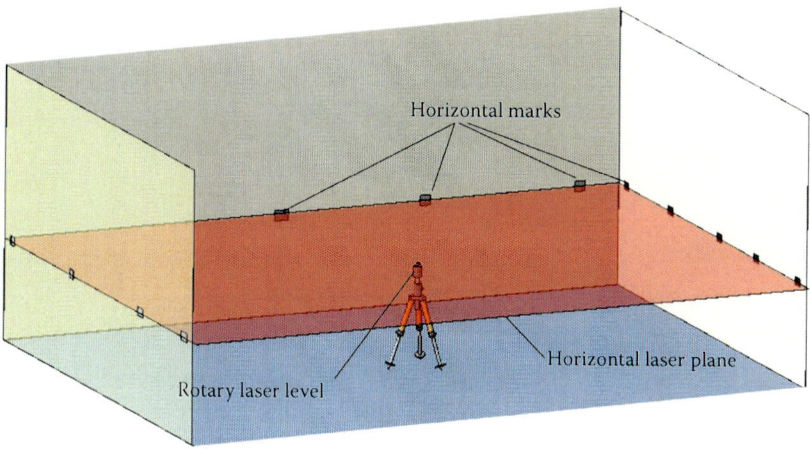

FIGURE 2.41 Checking and adjusting of a rotary laser level.

horizontal position with the help of a circular level. The circular level is usually set on the instrument's body or a tribrach. We incline the instrument and check its operational range in two perpendicular directions. If the instrument is greatly inclined the laser will start blinking. In more advanced instruments this blinking is followed by an audible alert. Asymmetry of the operating range for a self-leveling mechanism can be corrected by shifting the annular limiting contact.

During examination of a laser level with an inverse pendulum we can determine undercompensation or overcompensation the same way we did in an automatic optical level. We can reach the optimal operational state of the self-leveling mechanism by changing the weight of a counterbalance. In the case of overcompensation we should decrease the counterbalance weight, otherwise the weight should be increased.

Laser levels with a cardan suspension have special features. Effective operation of a cardan suspension can be affected by dirty bearings, bending axes, or gaps appearing after physical impact. We can confirm this by lightly shaking the instrument. The laser line must turn back to the marks. If it does not, the cardan suspension should be deemed defective.

BIBLIOGRAPHY

Books

Glendinning, J. and J. G. Olliver. 1969. *Principle and Use of* Surveying *Instruments*. London: Blackie & Son Limited.

Patents

Armstrong, J. A. 1961. Optical compensator and tilt detector. US Patent 2,981,141, filed January 16, 1958, and issued April 25, 1961.

Baker, A. L. 1965. Self compensating telescope level having fixed and pendulum mounted pairs of reflecting surfaces. US Patent 3,220,297 filed February 26, 1959, and issued November 30, 1965.

Bozzo, M. D. Device for projecting a flat beam of diverging laser rays. 1998. US Patent 5,782,003 filed February 21, 1996, and issued July 21, 1998.

Drodofsky, M. 1957. Device for determining small inclinations from the vertical or horizontal. US Patent 2,779,231, filed November 9, 1950, and issued January 29, 1957.

Gaechter, B., B. Braunker, and F. Muller. 1987. Measuring device for determining relative position between two items. US Patent 4,715,714 filed December 27, 1984, and issued December 29, 1987.

Hickerson, R. R. 1968. Surveyor's automatic level. US Patent 3,364,810 filed February 24, 1964, and issued January 23, 1968.

Kumagai, K., S. Kawashima, K. Furuya, and F. Ohtomo. 1998. Electronic leveling apparatus and leveling staff used with the same. US Patent 5,742,378 filed October 7, 1996, and issued April 21, 1998.

Nagao, T. Electronic level and leveling rod for use in electronic level. 1999. US Patent 5,887,354 filed April 10, 1997, and issued March 30, 1999.

Nakasawa, N. 1971. Optical instrument with stabilized reflecting component mounted as inverted pendulum. US Patent 3,582,179 filed June 19, 1968, and issued June 1, 1971.

Nau, K. R. 2008. Laser light receiver apparatus with automatically adjustable zero-reference point. US Patent 7,414,704 B1 filed July 25, 2007, and issued August 19, 2008.

Ohtomo, F. and S. Hirano. 1999. Laser survey instrument. US Patent 5,898,490 filed March 11, 1997, and issued April 27, 1999.

Qi, Y.-C. and H.-T. Liu. Laser level with improved leveling adjustability. US Patent 7,373,724 B2 filed February 14, 2006, and issued May 20, 2008.

Shirai, M., H. Takayama, and K. Kaneko. 2004. Surveying instrument having phase-difference detection type focus detecting device. US Patent 6,677,568 B2 filed May 10, 2001, and issued January 13, 2004.

Suzuki, S. and M. Nakata. 1999. Automatic focusing apparatus for detecting and correcting focus state if image optical system. US Patent 5,856,664 filed March 17, 1997, and issued January 5, 1999.

Wild, H. 1912. Spirit level with reflecting system. US Patent 1,034,049, filed August 2, 1910, and issued July 30, 1912.

3 Theodolites

A theodolite is a surveying instrument used for precise angular measurement in both horizontal and vertical planes. Theodolites are commonly used for land surveying, route surveying, construction surveying, and in the engineering industry.

3.1 HISTORICAL PROTOTYPES OF MODERN THEODOLITES

We can consider Heron of Alexandria's dioptra (first century BC) as the modern theodolite prototype. Prior to Heron's invention, ancient scientists applied primitive goniometrical instruments in astronomy and building. In astronomy, mainly vertical angles were measured, and only horizontal angles were measured in building. Heron's merit is invention of a universal goniometrical instrument (Figure 3.1).

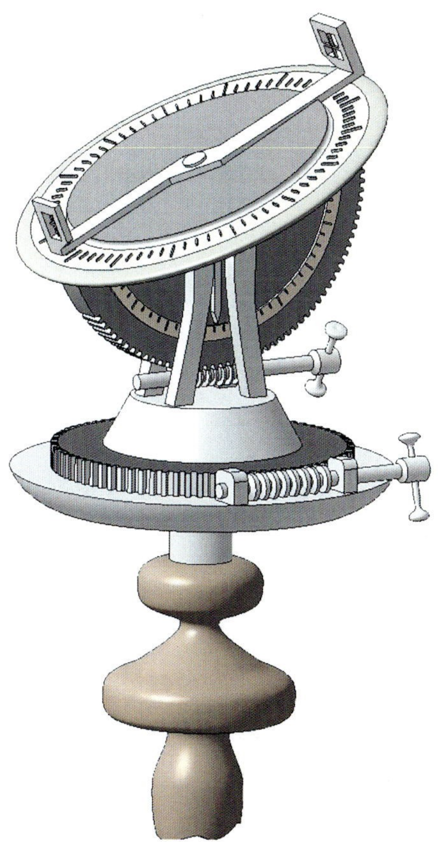

FIGURE 3.1 Heron's dioptra.

He also worked out methods for practical use of the instrument. Applying those methods people could carry out joining of water supply tunnels that they dug up from opposite sides of a mountain!

Over time goniometrical instruments eventually became equipped with a compass for orientation, a tubular level, and a Kepler telescope. In that time, the Kepler telescope could only provide external focusing. That meant the need to remove the ocular along the optical axis of the telescope. The term "theodolite" was introduced by Leonard Digges in the fourteenth century, but it only referred to an instrument that measured horizontal angles. The next significant step was the fitting of the theodolite with a telescope, made in 1725 by Jonathan Sisson. By the end of the nineteenth century the instrument looked like what we see in Figure 3.2. At that point the theodolite had metallic circles (limbs). Measuring was fulfilled by the means of two diametrically opposite microscopes. Therefore, eccentricity of the circles' influence was minimized. Presence of three or four lifting screws at the tribrach was the main feature. A precise tubular level was often placed on the Kepler telescope. The compass was an important instrument for orientation and was usually placed between the

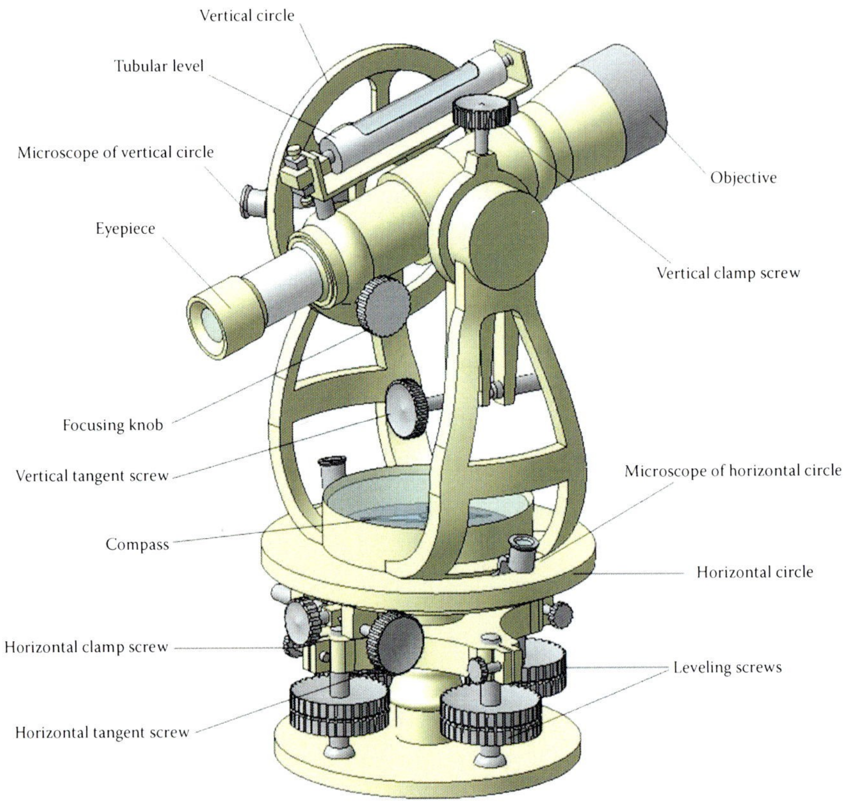

FIGURE 3.2 Theodolite with metallic circles (limbs).

standards. Fastening and focusing screws were separated, which may be present in modern elementary theodolites.

3.2 OPTICAL THEODOLITE

In the 1920s, leading surveying instrument manufacturers started using glass limbs in their theodolites. Nevertheless, metallic limbs were still applied in theodolites until the 1960s. About the same time with the glass limb-style theodolites appearance, another type of theodolite with an internal focusing telescope appeared. Instead of the compass, a tubular level was set up between the standards. The compass was moved onto the standard and it became demountable. The separate microscopes were replaced with the common one, and its ocular was set next to the telescope ocular. Also, an optical plummet was added. Replacement of the separate tubular level at the vertical circle with an optical and mechanical compensator was the last improvement of optical theodolites. The most advanced theodolites have coaxial fastening and focusing screws, instead of separate ones. The last improvements of the optical theodolites were carried out in the 1990s. An up-to-date optical theodolite is shown in Figure 3.3. Current surveying instrument manufacturers have stopped developing

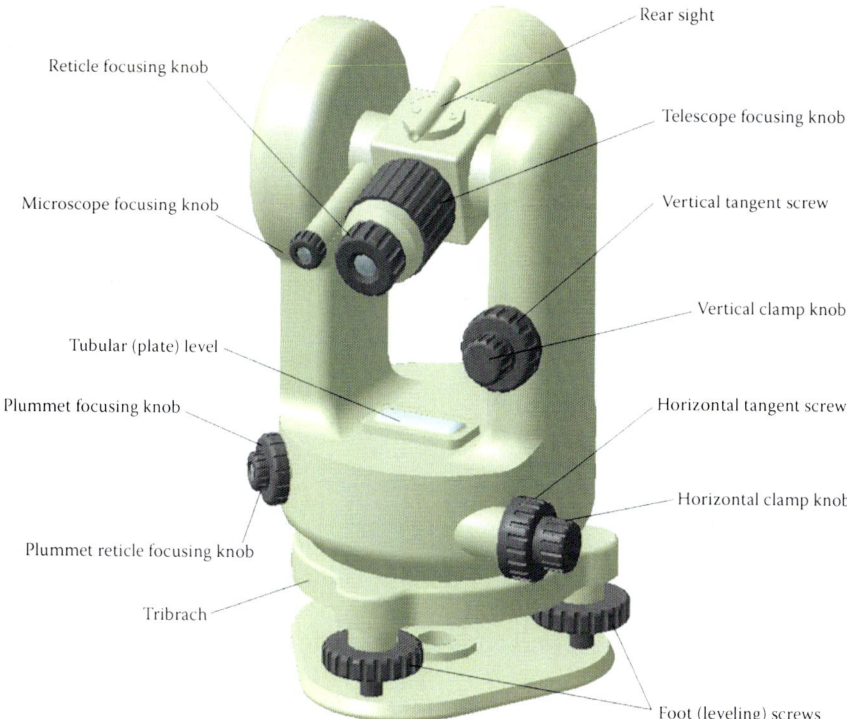

FIGURE 3.3 Optical theodolite.

TABLE 3.1
Up-to-Date Optical Theodolites

Model	Angle Measure Accuracy (″)	Magnification (*n*×)	Compensator Setting Accuracy/ Operating Range (*n*″/*n*′)	Tubular Level Accuracy (*n*″/2 mm)	Minimal Focusing Range (m)	Manufacturer
TD-1E	1	30	0.3/±2	20	2	Boif
TDJ2E	2	30	0.3/±2	20	2	Boif
TDJ6E	6	30	1.0/±2	30	2	Boif
ADA POF-X15	15	28	—	30	2	ADA Instruments
FET 500	30	20	—	30	1.2	Geo-Fennel

and releasing optical theodolites. However, some manufactures still make them available, mainly under licenses (Table 3.1).

3.3 ELECTRONIC THEODOLITE

At the peak of their development optical theodolites became reliable, compact, light, and ergonomic, but reading out the values remained tiring and hardly available for automatic registration. Some attempts were made to automate data registration in field conditions by taking photos of the limb parts at the moment of reading out. Then the film was processed in the lab and went into automatic counters. In the 1970s character recognition technology was poorly developed, so the values on the limbs were encoded with the help of white and black stripes. There is no doubt that the technologies of today would allow reading out the limb characters' image much more easily, but at that time people had to deal with various constraints. Thus, the first coded limbs on theodolites appeared. As electronic and microprocessor technology has progressed, it has become possible to fulfill the coded limb image processing technique in the theodolite. Such theodolites are called electronic theodolites. Nowadays, surveying instrument manufacturers produce them. An electronic theodolite has much in common with optical models (Figure 3.4).

The telescope, tribrach, optical plummet, focusing and fastening screws, and axes systems mainly remained unchanged. The measuring microscope disappeared due to lack of need. The digital display console with control keys appeared. Now there is a battery module at the right standard. The accuracy of many models released ranges from 2″ to 20″. Two-second accuracy theodolites have electronic monoaxial inclination compensators. Some of them even have a dual-axis compensator and a laser plummet. Five-second accuracy electronic theodolites usually include a monoaxial compensator. Some electronic theodolites are equipped with a laser pointer. Those of this type are called laser theodolites.

FIGURE 3.4 Electronic theodolite.

3.4 BASIC OPERATION PRINCIPLE OF A THEODOLITE

The main principle of every theodolite operation is a selected basic axial configuration according to certain requirements.

3.4.1 Basic Axes of a Theodolite

Optical and electronic theodolites have an identical geometric and kinematical scheme (Figure 3.5). This consists of vertical and horizontal rotation axes and the collimation axis. The vertical axis is the instrument rotation axis. The horizontal axis is the telescope rotation axis. The vertical rotation axis is provided with the horizontal measuring circle. The horizontal rotation axis is provided with the vertical measuring circle. These circles are often called limbs. The collimation axis is the line that connects the center of the telescope objective with the reticle's crosshairs.

3.4.1.1 Theodolite Vertical Axis

The vertical axis must be set into the plumb position at the beginning of a measurement. This is carried out by the means of the foot screws on the tribrach and using a tubular level as an indicator (Figure 3.6). Next we rotate the instrument and place the tubular level parallel to the line connecting foot screw 1 with foot screw 2. Then we set the bubble into the center of the tubular level turning foot screws 1 and 2. Next we turn the instrument at 90° around its vertical axis and again center the bubble by the means of foot screw 3. Then we turn the instrument 180° to check adjustment of the tubular level.

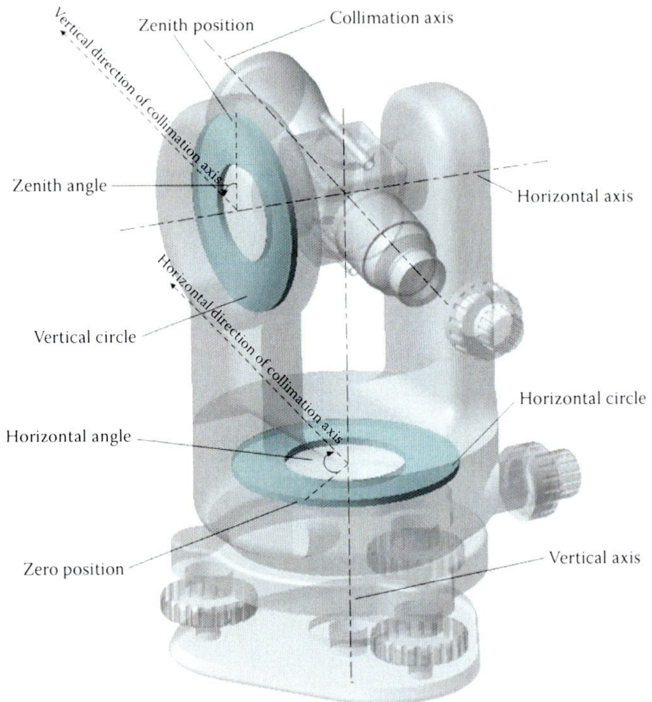

FIGURE 3.5 Basic axes of a theodolite.

If the bubble on the tubular level moves from the center, set it halfway back to the center by the means of leveling screw 3. Now we correct the other halfway by the means of the adjusting screw. We need to be sure that the bubble is in the center by rotating the instrument at 180°. If not, repeat the adjustment. We need to repeat checking and adjusting until the bubble is in the center at any instrument position.

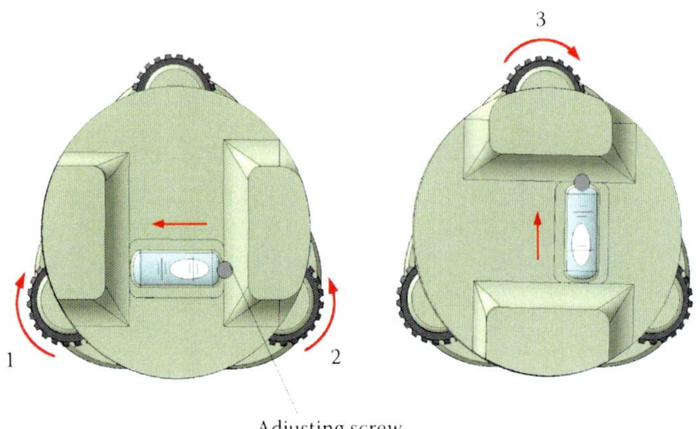

FIGURE 3.6 Tubular level adjusting.

The tubular level scale division ranges from 20″ to 60″ per 2 mm depending on the theodolite's precision. This allows us to establish the vertical axis accuracy from 10″ to 20″. This is enough for low precision theodolites. Moderate- and high-precision theodolites have monoaxial and dual-axial compensators for the instrument's vertical inclination for correct readings of vertical and horizontal angles.

It is an important requirement for the vertical axis to remain highly stable. When the instrument is new there is usually little worry about this, even with low-precision theodolites. However, after a shock or unskilled repair the tight vertical axis may develop some gaps or internal dents made by the bearing balls. The first sign of a problem is usually inadequate tubular level reactions during adjustment. In order to verify this malfunction we should direct our theodolite to a very clear target at a distance of about 10 m. Beforehand, we should set the instrument very carefully to the vertical position using a tubular level. Then, we unfasten the horizontal circle clamping screw and rotate the instrument several times to one direction and contrariwise. Before changing the rotation direction we should make sure that the horizontal line of the reticle and the target coincide. In the case of visible noncoincidence at any change of direction, and also attended with the bubble deviation, this indicates vertical axis instability. The problem is resolved by changing out the axial pair in a specialized workshop.

3.4.1.2 Theodolite Horizontal Axis

The horizontal axis must be perpendicular to the vertical. The horizontal axis is called the telescope rotation axis. The vertical axis is referred to as the instrument rotation axis. The horizontal axis nonperpendicularity to the vertical one is called the horizontal axis inclination.

Inclination of the horizontal axis ι distorts the horizontal circle reading out results at the value υ:

$$\upsilon = \iota \cdot tg\beta, \tag{3.1}$$

where β is the angle of telescope inclination (the vertical circle reading out).

The horizontal axis inclinations influence upon horizontal angle measurement values can be minimized by carrying out measurement at two circular positions (Figure 3.7).

One of the bushes of the horizontal axis can be slightly removed to regulate the axis inclination. The adjusting bush is placed in the standard that is without the vertical circle. Usually this is the right standard of the theodolite. Some manufacturers provide the option of regulation during the theodolite usage, while others rule out any access and set the bush with epoxy glue. Three commonly used types of a regulated bush fixations are in Figures 3.8 through 3.10.

The first type of fastening is the handiest. It is applied in Nikon, Trimble, Spectra Precision, and Pentax instruments. Adjustment is fulfilled by the means of two screws that have conical tips. Before adjusting, the flange fastening screws are slightly loosened. We need to remove the battery and open the rubber plugs to reach these screws.

The regulating screws may also be covered with rubber plugs. While rotating the adjusting screws in any direction we can rotate the bearing flange at a slight angle around the pin. The horizontal axis is slightly removed at height. After adjusting we should tighten the fastening screws.

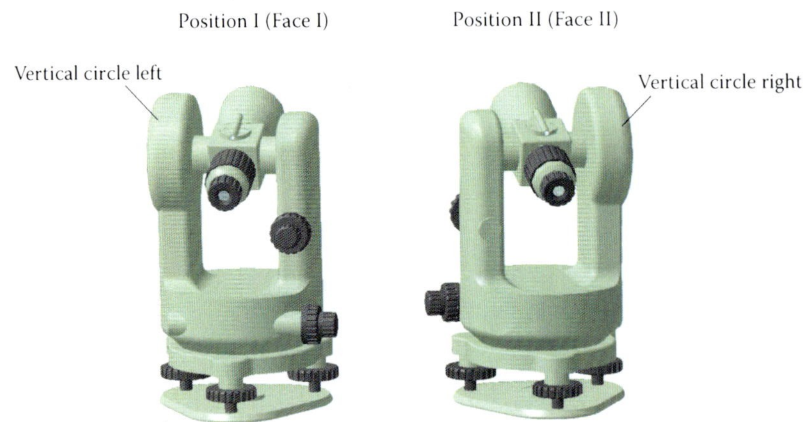

FIGURE 3.7 Theodolite positions.

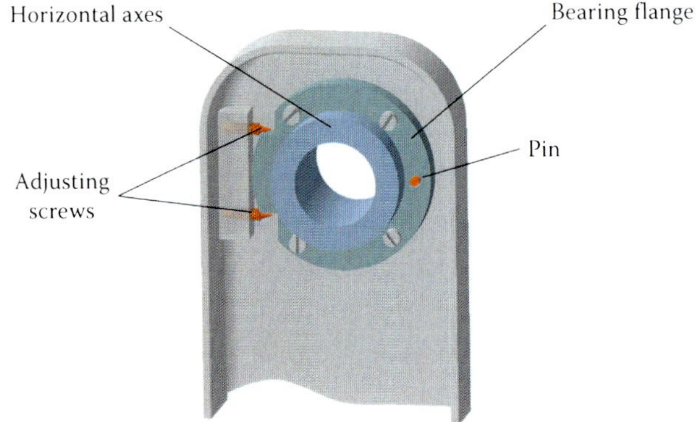

FIGURE 3.8 Units for Nikon theodolite horizontal axis inclination adjustment.

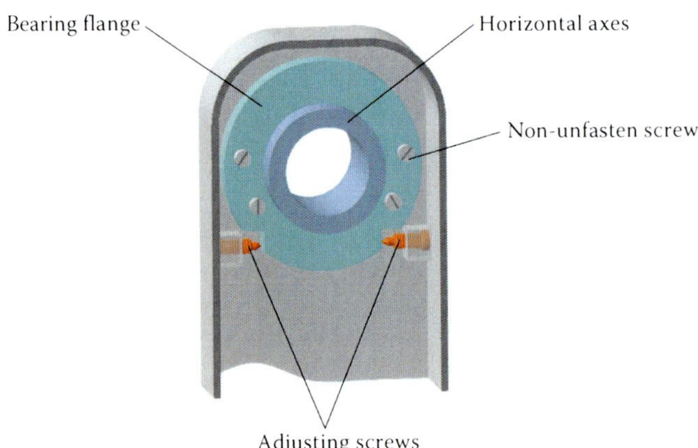

FIGURE 3.9 Units for Topcon theodolite horizontal axis inclination adjustment.

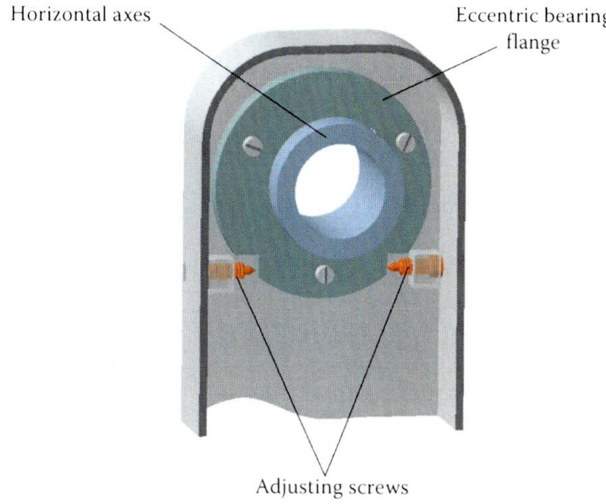

FIGURE 3.10 Units for Geo-Fennel theodolite horizontal axis inclination adjustment.

The second type is often applied in Topcon instruments. The main difference of this type is the lack of a pin. One of the lateral fastening screws is used as the pin. It is not loosened before adjusting. Another difference is that the adjusting screws are rotated in the same direction. The adjusting screws have spherical tips.

The third type is often applied in low-precision theodolites. The horizontal axis removal is fulfilled by rotating the eccentric bearing flange by means of the adjusting screws.

If a theodolite has no horizontal axis inclination adjusting unit, then we can make slight adjustments via the fastening screws on the vertical axis flange.

These screws are placed between the theodolite standards and protected with a cover or rubber plugs. Adjusting is fulfilled by means of lateral fastening screws (Figure 3.11). We can do it only by tightening one of the screws at the required side,

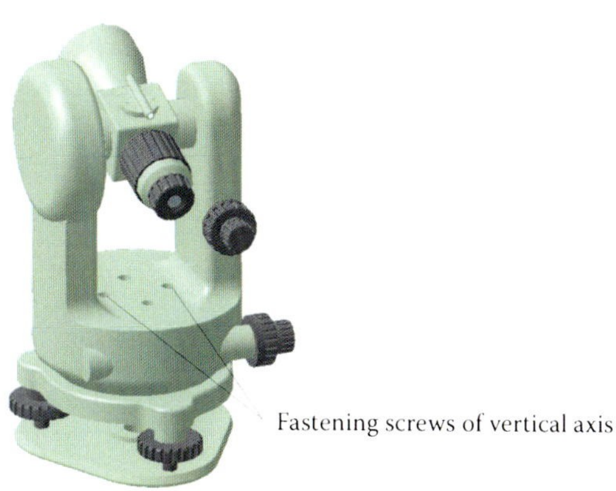

FIGURE 3.11 The alternative method for elimination of theodolite horizontal axis inclination.

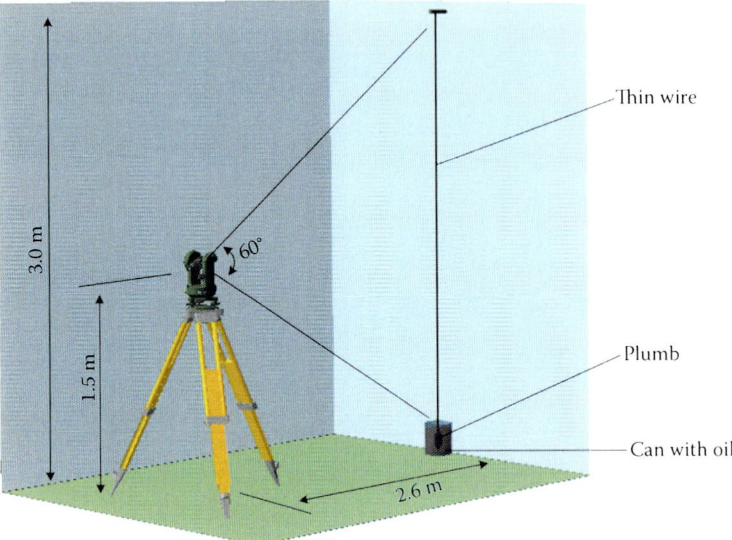

FIGURE 3.12 Theodolite horizontal axis inclination checking.

and never loosening the opposite screw. This method is not very efficient because after adjusting we should adjust the compensator.

Next we make a fundamental equipment assessment of the theodolite axes perpendicularity. We can examine this in two ways. The first way is shown in Figure 3.12. Set up the theodolite on the tripod at the distance of 2.6 m from the wall. A thin wire with a weight is suspended from the top of the wall. In order to remove oscillations of the wire, the weight is put into a can with oil.

The wire thickness should be about 0.1 mm. Its angular size is 5″ at the distance of 3 m from the theodolite objective. We can use the horizontal theodolite circle or the reticle bisector to measure small angles. The reticle bisector angular size depends on the theodolites precision and can be equal to 20″, 30″, 40″, or 60″.

The second method uses a mark and a ruler, graduated in millimeters. The mark is placed at the top of the wall. The ruler is placed horizontally at the bottom. The ruler must have thin and clear-cut lines. The angular size of 1 mm division at the same distance of 3 m is about 50″. This is sufficient enough for low- and moderate-precision theodolites adjustment.

We test the horizontal axis inclination in the following way. Direct the telescope to the upper ending of the wire (or to the mark) at one of the circle positions. Then unfasten the vertical clamp and direct the telescope to the lower ending of the wire (or to the ruler). The vertical line of the reticle may coincide slightly with the center of the wire. This is natural because some inclination of the vertical axis may occur. Then we find out the deviation by the means of the reticle bisector or the horizontal circle of the theodolite. If we apply the second method, we should do the ruler reading. Then we turn the theodolite to another position and direct it again to the upper target. Again we direct it to the lower target. The vertical direction deviation from the lower target at both positions of the theodolite should not be more than 10″ for moderate- and high-precision theodolites. The 30″ difference is allowable for low-precision theodolites. In case we try the second way, the rulers' read out difference should not exceed 0.2 mm (0.6 mm for low-precision theodolites). If the limits are exceeded, we should correct the horizontal axis inclination with the aid of accommodation described earlier or the vertical axis flange fastening screws.

3.4.1.3 Theodolite Collimation Axis

The collimation axis of the telescope should be perpendicular to the horizontal axis of the theodolite. Nonperpendicularity of these axes is called collimation error *C* and influences the horizontal angle read out value ε like this:

$$\varepsilon = \frac{C}{\cos\beta} \tag{3.2}$$

where β is the angle of telescope inclination.

The collimation error's influence on horizontal angle readings could be excluded the following way. The horizontal angle's measurements are fulfilled at two positions of the theodolite and then the result is averaged. Surely we should take into account the 180° difference between two positions at the same direction. The double collimation error is the angular read out difference from 180° at the same direction for both positions of the theodolite. The collimation error should not exceed 10″ for high-precision theodolites. It need to be less than 20″ for moderate-precision theodolites and not exceed 60″ for low-precision theodolites. In case it exceeds these values we should adjust the instrument with of the horizontal adjusting screws on the reticle (see Figure 3.13).

Before collimation error corrections, we should be sure that reticle inclination has not occurred. It is convenient to use a suspended vertical wire (see Figure 3.12). First, we should properly set the vertical axis of the theodolite into the vertical position. In case the wire image does not coincide with the vertical line of the reticle, we should slightly loosen the ocular flange's fastening screws and turn the flange at the required angle. Then we tighten the screws. There is another suggested method for reticle inclination adjustment. We start by superposing the reticle vertical line with the target. Then we remove the target image to the lower edge of the reticle by means

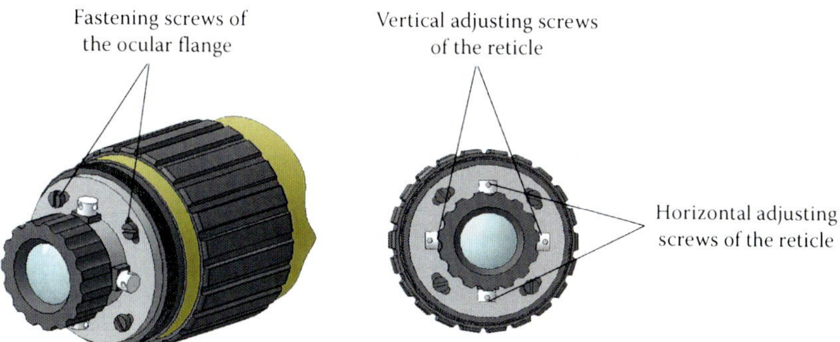

FIGURE 3.13 Reticle adjusting screws.

of the vertical tangent screw. In the case that the image removes more than the size line thickness, adjusting is needed.

The collimation axis of the telescope should be horizontal when the vertical circle read out is equal to zero. In order to meet this requirement, we should measure the vertical angle at two positions of the theodolite. The total sum of these readings must be 360° if the theodolite has an ordinary full scale (from 0° to 360°) of the vertical circle. Some low-precision theodolites have an inclination scale of ±90° instead of the full scale. In this case, sightings of the same target should have angles of inclination at both positions of the theodolites and must be equal but have opposite signs. The difference of the sum from 360° (0° for the instruments with the inclination scale) divided in two is called the vertical circle index error. In order to correct it we should correct the vertical circle read out by means of the vertical tangent screw. Then we superpose the horizontal line of the reticle to the target with the help of the vertical adjusting screws (see Figure 3.13). We suggest correcting only slight vertical index errors with the help of these screws. If the vertical index value is several minutes, the reticle's horizontal removal or inclination could appear. The horizontal removal of the reticle changes the collimation error value that must be corrected. The vertical index adjustment of low-precision theodolites can be fulfilled only adjusting the reticle screws.

Optical theodolites equipped with an inclination compensator of the vertical axis usually have options for regulating the vertical index via compensator adjustments.

All electronic theodolites have special programs to calculate the vertical index error. Users are advised to use the correcting program instead of using the vertical adjusting screws of the reticle. The program is usually initiated by a simultaneous keypress combination (which is specific for every manufacturer and described in their manuals) or entering a special menu. Then we usually point at the target twice from different theodolite positions. After each sighting we should press the Enter key. After the second input the vertical index error correction is fulfilled automatically. Electronic theodolites without a compensator are adjusted in such a way without any troubles. A more complicated adjustment is needed for electronic theodolites with a compensator.

If an electronic theodolite experiences an impact, software-based vertical index adjustments may be incorrect. This occurs due to compensator shift after the shock. In order to verify the vertical index position we should put the telescope into the horizontal position by setting the vertical circle read out equal to 90° (or 0°). Then we test it like a usual optical level with the leveling rods.

3.5 MAIN PARTS OF A THEODOLITE

3.5.1 Measuring System of a Theodolite

3.5.1.1 Measuring System of an Optical Theodolite

An optical theodolite measuring system consists of horizontal and vertical glass limbs, plus reading units. Optical theodolites glass transparent limbs have circular scales graduating from 10′ to 1°. Degree divisions are added with Arabic figures. The optical theodolite reading device is a microscope furnished with an index or scale micrometer.

The measuring system of an elementary contemporary optical theodolite is shown in Figure 3.14. The outside light illuminates the vertical limb through the

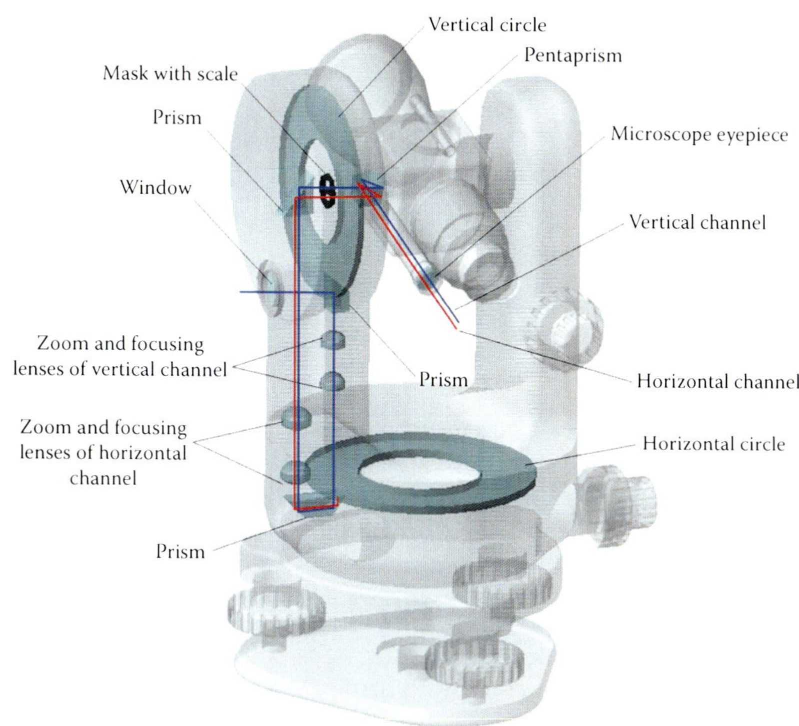

FIGURE 3.14 Optical theodolite measuring system.

matte window. Then the light goes through the right-angled prism of the vertical channel and comes to the transparent horizontal limb. The horizontal and vertical scale images do not overlap each other and are parallel if adjustment is correct. Then the images enter the horizontal microscope. As a matter of fact, it is common for both vertical and horizontal channels. This is why after the horizontal channel image adjustment we must confirm the vertical channel image. The optical scheme of this kind is called consecutive. Having gone through the microscope, the circles' images enter the right-angled prism, which sends the images to the mask. The microscope mask is like the telescope reticle. It has two separate transparent windows for the vertical and horizontal channels. Various types of microscopes have different windows. Elementary microscopes have index-marked windows (see Figure 3.15). The microscopes of moderate precision theodolites have scaled windows.

The vertical and horizontal circles' images superposed with the mask enter the pentaprism and then the microscope's ocular.

3.5.1.2 Measuring System of an Electronic Theodolite

Electronic theodolites limbs are covered with a nontransparent coating that has code gaps on it. They may have regular intervals (incremental solution) and irregular ones (barcode solution). A five-photodiode matrix is used as a reader in incremental solution. The CCD (charge-coupled device) line is applied as a reader in barcode solution.

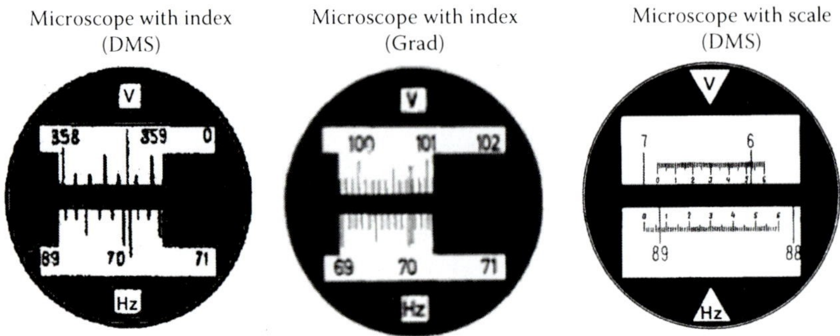

FIGURE 3.15 Reading eyepieces fields of view.

An electronic theodolite incremental measuring system is a sort of accumulative measuring system. Before measuring they are forcedly zeroed. While measuring the incremental system accumulates small parts of the measured quantity. A classic example of these units is a clock. An ordinary clock is an incremental quantity irreversible system to measure time. A photoelectronic incremental irreversible system for distance measuring is in the top part of Figure 3.16.

The source of light (light-emitting diode) is formed into a narrow beam with a condenser lens and a mask that has a slit. A slit grid is set in front of the photodetector. At the moment the slit grid moves, sinusoidal modulation of light occurs at the photodetector input. Monochannel irreversible solutions are seldom used. In the bottom part of Figure 3.16 there are two channels, which is necessary to provide reversibility. Since distance is able to increase or, on the contrary decrease, in practice only a dual-channel reversible system is used to measure distance. The sensor has two slits shifted one relative to the other at a one-fourth period phase of the grid step. There are also two photodetectors. When the grid moves to one direction a sinusoidal signal at one of the photodiodes output advances the signal at the output of the other photodiode. When the grid moves in the reverse direction the signals sequence is opposite.

Angular measuring incremental systems are based upon the same principle. The slit grid is set circle-wise, and the angle is identified as the distance passed by the slit mask around the circle. There are several tenths of slits on the mask to increase signals at the photodiodes outputs. The mask slits are distributed at the same step as the grid spacing around the limb.

An incremental measuring system of an electronic theodolite is in Figure 3.17. The incremental limb scale is a regular sequence of equal dark and transparent stripes. The angular interval between them is from 1′ to 2′. The limb also has a short barcode strip for zeroing. There is an immovable mask at a very small distance (from 5 to 10 mkm) from the scales (Figure 3.18).

There is a light source at one side of the limb and a five-photodiode matrix at another one. The mask is made nontransparent, but it has five transparent stencils. One of them has a barcode strip identical to that the limb has. When we rotate the limb once their complete superposition occurs, and the zero photodiode generates a short impulse. The other four stencils consist of sequences of transparent stripes with

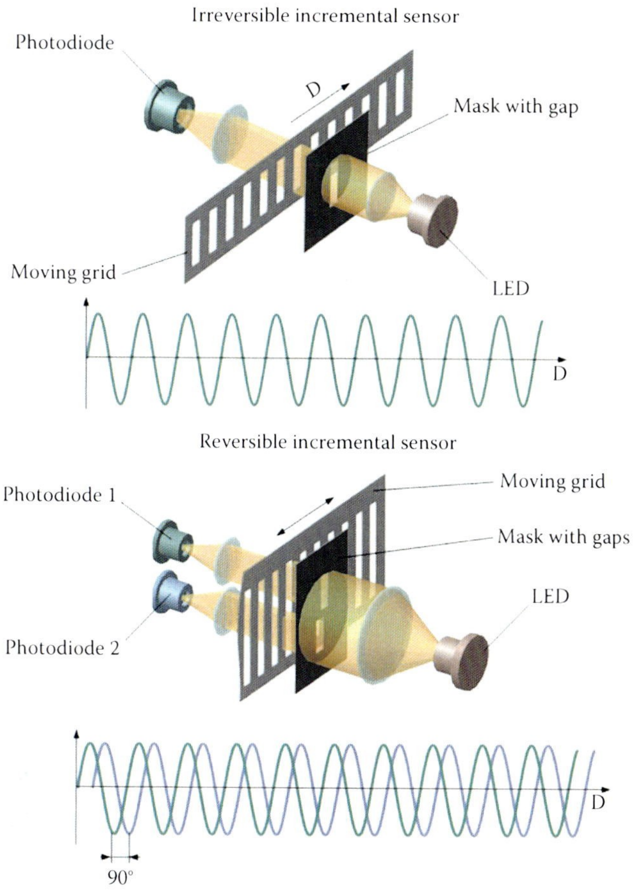

FIGURE 3.16 Incremental measuring principle.

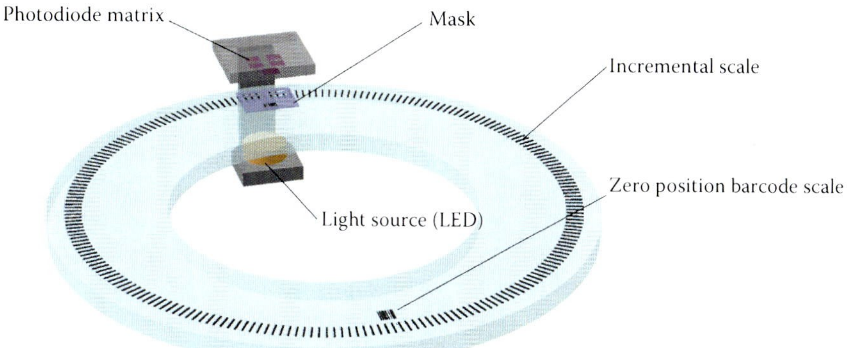

FIGURE 3.17 Incremental measuring system.

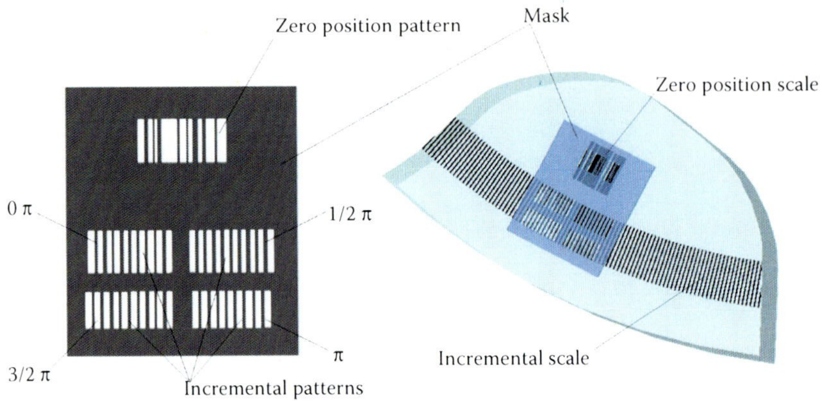

FIGURE 3.18 Mask and incremental scale.

the same periods as on the limb. However, these stencils are shifted at one-fourth of the period from each other. While rotating the limb, four sinusoidal signals are generated at the corresponding photodiodes' outputs. The phase shift of these signals is 90°. Further, these signals are processed by the means of two units: a reverse counter and an interpolator. Before entering the reverse counter, the sinusoidal signals are transformed into impulse ones. Further, the pairs of 90° phase shift signals are analyzed. While rotating the limb at one direction, the first pair of impulses advances the second pair of impulses. When we change the direction of the limb rotation, the impulse sequence is changed too. These impulses enter the trigger, which is sensitive to changes of these signals' sequences at its inputs. The trigger is switched over at every change of the limb rotation direction. The trigger controls a reverse impulse counter. The impulse sequence from one of four channels enters the counter input. The data accumulated by the reverse counter is equal to the current angular value. The value discreteness is from 1′ to 2′. More exact angular value can be attained by the means of the interpolator. It carries out preliminary analog processing of the sinusoidal signals and then they enter an analog-to-digital converter. The preliminary analog-digital processing is necessary to minimize the constant signal drift. This is why 180° phase shift signals are processed in pairs. Data from both the reverse counter and the analog-to-digital converter enter the theodolite microprocessor. Using the data the microprocessor calculates the angular value to within 1″.

The incremental angular measuring system was most widespread 10–20 years ago. At that time all leading manufacturers, except Leica, created electronic theodolites on this principle. Nowadays, this principle is slowly excluded by more advanced absolute methods. Today only a quarter of electronic theodolites made use incremental sensors.

The absolute method is based on the fact that any position of a limb corresponds to an appointed angular value. Optical theodolite measuring systems are similar to absolute systems. Electronic theodolites have absolute code limbs (Figure 3.19).

There are several types of limb coding. In the past there were multitrack code limbs in angular measuring surveying instruments. Because of CCD line technology

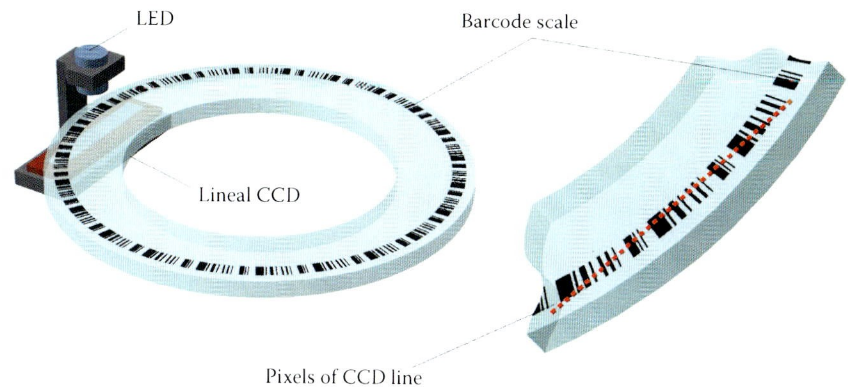

FIGURE 3.19 Barcode measuring system.

development, nowadays only barcode solutions are used in absolute electronic theodolites. Such a limb has an infinite barcode stripe circumferentially spaced. An absolute angular sensor consists of a light-emitting diode and the CCD line where the barcode stripe images are projected. The CCD signal is processed the same way as it was described in Chapter 2 about digital levels. The only difference is that a digital rod is coded in linear values, whereas a barcode limb is coded in angular values. This is the same way we find out the exact part of the angular value according to the phase shift of the barcode support grid. This is how we find millimeters and their fractions in a digital level. There are several systems of limb coding. Usually they are unified by every manufacturer. For example, Topcon applies the same phase-measuring method to code leveling rods and their theodolites limbs. Other leading manufacturers use their technical groundwork both in digital theodolites and levels.

3.5.1.3 Influence of Wrong Limb Position on Angular Measuring System Accuracy

Theodolite measuring system errors may occur because of wrong positions for either the limbs or sensors. The errors occur if the center of the limb scale is not on the axis of rotation and also if the limb plane is inclined to this axis (Figure 3.20). Such errors are called limb eccentricity and limb tilting.

Limb eccentricity is one of the main reasons for theodolites measurement errors, and it can hardly be corrected. Let us analyze the eccentricity formula:

$$\beta = \left(\frac{l}{r}\right)\rho'' \sin\alpha \tag{3.3}$$

where β is the eccentricity influence on the angular read out, l is the linear component of eccentricity, r is the limb radius, ρ'' equals 206265″, and α is the angular component of eccentricity.

We take a typical limb of an 80 mm diameter and then superpose it with the axis of rotation. Usually the superposition accuracy is from 1 to 2 mkm. According

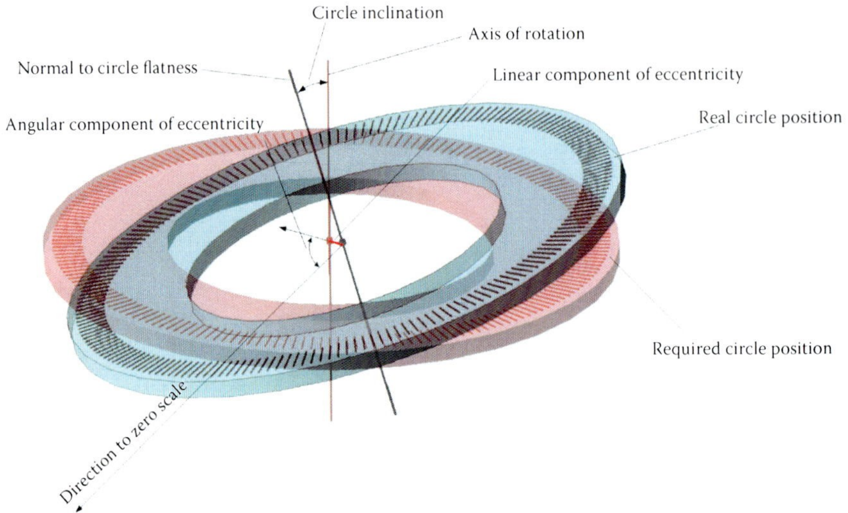

FIGURE 3.20 Limb position errors.

to this formula we evaluate the maximal value angular error as from 5″ to 10″! Now we realize that we need not only high accuracy of theodolite adjustment, but also the highest quality of axial systems and bearings. The eccentricity influence could be minimized methodically by measuring the angle at two positions of the theodolite (see Figure 3.7). Two diametrically opposed sensors are set in high- and moderate-precision electronic theodolites to minimize this error. Some of leading manufacturers apply mathematical correction methods in electronic theodolites. After assemblage, the instrument is tested on an angle-measuring stand. According to testing, angular and linear components of this error are determined. Then they are written into the permanent storage of the microprocessor, which calculates eccentricity correction data and inserts them into every angular reading out.

In optical theodolites significant values of limb eccentricities may be visible. We could see limb images shifting about the mask edges while rotating the theodolite. It is advised to test limb eccentricity influence in the lab. In the center of a room with a stable floor we set up our tested theodolite. In order to test the horizontal circle eccentricity, we put from six to twelve marks at the same angular interval on the walls of the room. The marks must be set up at the same horizontal line and it is advisable that they be at the same distance from the theodolite. Then we carry out angular measurements pointing to these marks at both positions of the theodolite. Now we calculate collimating errors of every direction. Then we draw a diagram illustrating collimation error dependence of the horizontal limb position (Figure 3.21).

The diagram has a sinusoidal configuration, especially when errors are significant. The diagram amplitude should not exceed an acceptable collimating error for definite rating of a theodolites precision.

If we do not have the opportunity to distribute the marks evenly along the horizontal line, we may place only four or three marks distributed evenly within the angle of about 100°. Then we outline the tribrach position on the tripod base with a

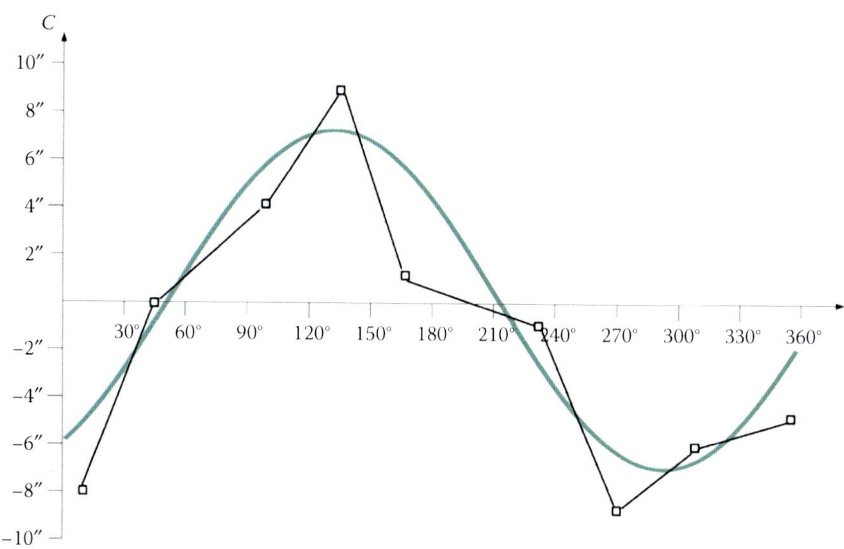

FIGURE 3.21 Limb eccentricity diagram.

pointed pencil. The next step it so measure the angles pointing to the marks at two theodolite positions. Then carefully unfasten the fastening screw of the tripod base and turn the theodolite at an angle of 120°. Then we superpose the tribrach with the contour on the tripod base and secure the tripod fastening screw. If we are testing an electronic incremental theodolite, we should not power it off during the test. Now fulfill the measurements again by pointing to the marks at two positions of the theodolite. Again we rearrange the instrument at 120° and fulfill the same measurements pointing to the marks. In this manner, we have from nine to twelve directions to test the horizontal circle eccentricity.

The vertical circle eccentricity test is less difficult. We should test out the eccentricity influence only during the operating range of the vertical circle ±30°. Three marks will do. One of them is set near to the horizontal line and two others are placed on the operating range edges. One of the marks is placed at the angle of 30° above the horizontal line and the other at the same angle below the horizontal line. The vertical angular measurements are carried out at two positions of the theodolite pointing to these marks. Then we calculate the zero positions (vertical indexes) for three vertical directions. If the zero positions are the same, eccentricity does not exist. If eccentricity exists, the zero position should not exceed the utmost limits for theodolites of this type of accuracy.

Limb inclination has very little geometrical influence on angular reads. Even an inclination of several minutes does not influence the result. Nevertheless, the limb inclination value must be less than one angular minute because of the following reasons. In an optical theodolite, modification of the distance between the microscope and the limb may be the reason for the limb image defocusing at various sections. In an electronic theodolite, this change of distance may have resulted in the sensor's malfunction because of the signal changing level. The limb inclination is especially unsafe for incremental electronic theodolites. The incremental sensor mask is usually set up at the distance of 10 mkm from the limb; that is why the limb inclination may have led to the mask and the limb coming in contact with each other. In this case they may be destroyed.

It is known that theodolites produced by leading manufacturers have properly setup limbs. Collisions seldom occur while using the theodolites, for their horizontal circles are properly protected and have strong axes. Meanwhile, the vertical limb may have changed its position in case of physical shock. The telescope is especially sensitive to impacts. Any time a theodolite is dropped we should test the limbs eccentricities.

3.5.2 Vertical Index Compensator of a Theodolite

3.5.2.1 Vertical Index Compensator of an Optical Theodolite

Moderate precision optical theodolites have a more complicated optical scheme of the vertical channel because of existence of the compensator's vertical circle index (Figure 3.22).

Compensation is carried out the following way. A parallel glass plate suspended on elastic strips is set between the vertical circle microscope and a mask. The suspension scheme is similar to a level compensator with a reverse pendulum. It is balanced by adjusting weights located at the upper part of the compensator pendulum. When the theodolite vertical axis is inclined along the longitudinal direction x the parallel plate glass rotates around its axis keeping its previous balanced state. During this rotation, the vertical limb image is shifted about the mask scale at the required compensation value. At the moment of theodolite vertical axis inclination, the transversal direction y of compensation does not occur. Therefore, when using optical theodolites it is necessary to closely watch the bubble position in the tubular level. In a moderate precision theodolite the level is always set in a transversal position.

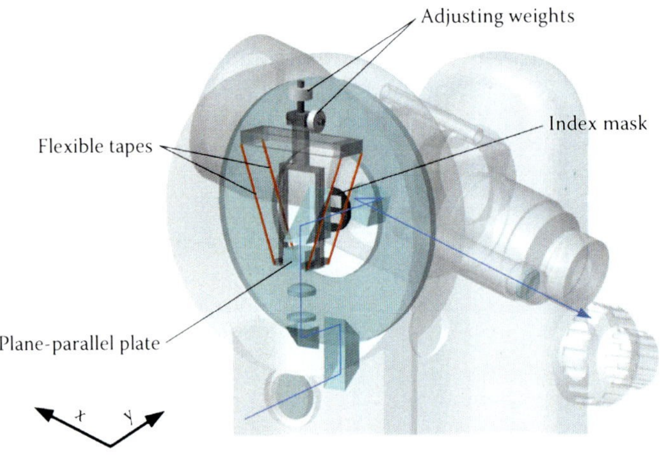

FIGURE 3.22 Vertical index compensator of an optical theodolite.

3.5.2.2 Vertical Index Compensator of an Electronic Theodolite

Compensators in electronic theodolites fulfill the same function that they do in optical theodolites, that is, they minimize influence of the vertical axis inclination upon the measurement results. Nevertheless, this problem in an electronic theodolite is solved in different manner than in their optical counterparts. An optical theodolite compensator changes the beam's motion in the optical readout unit. The beam's motion depends on the vertical axis inclination.

In electronic theodolites, the compensator is an independent device that measures slight angular inclinations of the vertical axis. Data from the compensator enters the microprocessor of the theodolite. It is up to us what to do with the data. We are able to give instructions to the microprocessor to take the data into account of angular measuring results. We can switch off the compensator or put the data into the display for estimating the instruments inclination. There are electronic theodolites that do not have a tubular level. In this case, we can use an electronic level to set the vertical axis into the plumb line. The theodolite preliminary setting is carried out with the circular level. The inclination along the direction x mostly influences the measurement results. The direction is parallel to the plane of rotation of the telescope. The vertical axis inclination to the x direction directly influences the vertical angle measurement result. The y direction is perpendicular to the x direction. So we can see from Equation 2.1 that the vertical axis inclination along the y direction has less of an influence on the measurement results. That is why dual-axial compensators are usually applied in a total station and seldom in a theodolite.

A monoaxial compensator is applied in electronic theodolites where accuracy is 5″ and above (Table 3.2). Unfortunately, some manufacturers do not set up compensators into 5″ accuracy theodolites. This seems to show that it does not concern leading manufacturers. For instance, even Leica sets up dual-axial compensators into 9″ accuracy theodolites.

Now we will look at a typical monoaxial compensator that is set up in most electronic theodolites (Figure 3.23).

The compensator's main component is a tubular fluidal level whose external side has some metallic contacts. They are used as variable capacitor plates. Operation of such a tubular capacitor leveling cell was mentioned in Chapter 2. We should point out that the compensator in an electronic theodolite must meet higher requirements.

We know that the length of the bubble in a tubular level depends on temperature. Tubular levels with vials whose sensitivities are from 20″ per 2 mm to 30″ per 2 mm are used in theodolites. Precision of the compensator provided with such tubular levels is about several seconds. This kind of precision in the whole operating range may be achieved only by taking into account temperature correction. That is why an electronic temperature sensor is set next to the vial. Data from the sensor enters directly into the theodolite microprocessor.

Any capacitor measurement system is very sensitive to electrical induction. That is why the compensator vial is protected with a metallic electrostatic screen.

On the bottom of the compensator bracket there are two holes to fasten it to the internal side of the theodolite standard. If we need to adjust the compensator, we should slightly loosen the fastening screws into these holes. Through gentle tapping

TABLE 3.2
Up-to-Date Electronic Theodolites Provided with Monoaxial Compensators (or No Compensator)

Model	Angle Measure Accuracy (″)	Magnification ($n\times$)	Compensator Working Range ($\pm n'$)	Tubular Level Accuracy (n''/2 m)	Minimal Focusing Range (m)	Manufacturer
DT202	2	30	3	30	0.9	Topcon
DT402	2	30	3	30	1	FOIF
DJD2-E	2	30	3	30	1.3	BOIF
ETH-302	2	30	3	30	1.35	Pentax
DET-2	2	30	3	30	1.35	Spectra Precision
ET-02	2	30	3	30	1.4	South
NE103	5	30	3	30	0.7	Nikon
DT205	5	30	3	40	0.9	Topcon
DT405	5	30	3	30	1	FOIF
ETH-305	5	30	3	30	1.35	Pentax
ET-05	5	30	3	30	1.4	South
DJD5-E	5	30	—[a]	30	1.3	BOIF
NE101	7	30	—	40	0.7	Nikon
DT207	7	30	—	40	0.9	Topcon
DT209	9	26	—	60	0.9	Topcon
NE100	10	30	—	60	0.7	Nikon
DJD10-E	10	30	—	30	1.3	BOIF
ETH310	10	30	—	40	1.35	Pentax
ETH320	20	30	—	40	1.35	Pentax
DJD20-E	20	30		30	1.3	BOIF

[a] — no compensator.

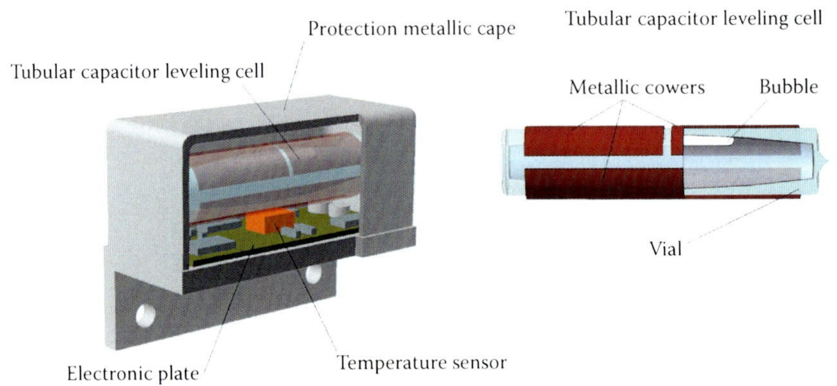

FIGURE 3.23 Monoaxial electronic compensator arrangement.

we can incline the compensator along the x direction until the tubular level axis is perpendicular to the vertical axis of rotation of the instrument. Afterward, the fastening screws should be tightened. As usual this kind of adjustment is initially set by the manufacturer, and if the theodolite is not disturbed, then the manufacturer adjustment will be sufficient during the service life.

As a rule, periodic electronic adjustments will suffice. Every electronic theodolite has special software for detecting the vertical circle zero position. The software is usually combined with an electronic level-adjusting program. Sometimes the electronic level-adjusting program is a detached point in the theodolite menu. More often, dual-axial compensators have such a program solution. All of these programs are available for users.

In case a theodolite has gone through a strong impact it is suggested to test the compensator. We must do this even if the theodolite correctly carries out program adjustment. During the test we should determine the compensator operating range and linearity of its work. We start by placing the theodolite at several meters from the wall so that one of the foot screws is directed to the wall. Now we set the vertical axis into the plumb position using the tubular level. Then we set the telescope horizontally, rotating it until the vertical angle reading out is equal to 0° or 90°. Then we look up the compensator's operating range in the technical specifications of the theodolite. It usually is 3′. Then we mark three index lines on the wall. One of them is horizontal, and other two are 3′ above and below the horizontal line correspondingly. Marking of these lines is fulfilled with the help of the vertical angle readout. The stand is now ready. Then we point the theodolite to the horizontal index line on the wall. Now we will turn the foot screw of the tribrach and superpose the horizontal line of the reticle with the upper index line on the wall. This way we incline the theodolite vertical axis at the 3′ level. Then write down the vertical angle value. Ideally it must be equal to 3′. The allowable difference is ±3″ for high-precision theodolites and it is ±5″ for moderate-precision theodolites. We test the compensator the same way inclining it in the opposite direction. At this point we superpose the reticle with the lower index line by the means the foot screw. If the deviations exceed the aforementioned values but are still the same at opposite inclinations, we can arrive at the conclusion of a nonrelevant scale factor.

If these deviations are asymmetrical that means the compensator shifted. The compensator position correction should be completed at a specialized workshop.

If you are well experienced in adjusting surveying instruments, you could try to adjust a monoaxial compensator by yourself. We would use the same stand. First, we set the vertical axis into the plumb position. Then we loosen the compensator slightly from the fastening screws. Then we put the telescope into the horizontal position and point to the horizontal index line on the wall. Now carefully turn the foot screw until the vertical angle readout stops changing. We mark this position on the wall. For the next step, we turn the foot screw to the opposite direction and mark the opposite point at which the compensator ceases operation.

Now we find the middle between these two points with a graduated millimeter ruler. Next we rotate the telescope and superpose the reticle with the middle. The vertical circle readout will now be different than 90° 00′00″. With light tapping on the compensator bracket, we try to get the readout close to 90° 00′00″. Twenty-second

TABLE 3.3
Theodolites Provided with Dual-Axes Compensators

Model	Angle Measure Accuracy (″)	Magnification ($n\times$)	Compensator Working Range ($\pm n'$)	Tubular Level Accuracy (n''/2 m)	Minimal Focusing Range (m)	Manufacturer
TM6100A	0.5	43	2	—	0.6	Leica Geosystems
DT210	2	30	3	30	0.9	Sokkia
DT510	5	30	3	40	0.9	Sokkia
2T5E	5	30	3	30	1	UOMZ
Builder T106	6		4	—	1.3	Leica Geosystems
Builder T109	9		4	—	1.3	Leica Geosystems

accuracy will do. We should not tap heavily, as the fragile vial of the level might crack. Now let us carefully tighten the compensator fastening screws. Afterward, we must complete the adjustment with the help of the compensator software and carry out the tests again.

Electronic theodolites with dual-axial compensators are seldom used. Some examples of this theodolites type are listed in Table 3.3. One of the best known dual-axial compensators is shown in Figure 3.24. It has often been used in total stations from leading manufacturers and is also set up in the Sokkia electronic theodolites. The main component of this compensator is a precise circular level. Its bottom is made of smooth optical glass. The source of light is set up below. Beams freely go through the bubble center of the circular level. The beams that reach the bubble edges are reflected and dispersed. Those beams that have passed through the bubble go up past the vial freely with minimal deflection to the center. If we set up a screen above the level, we can then see the annular shadow moving during the circular level's inclination. If we set up a four-photodiode matrix instead of the screen, we can watch the bubble's movement analyzing the photodiodes' signals. These photodiodes

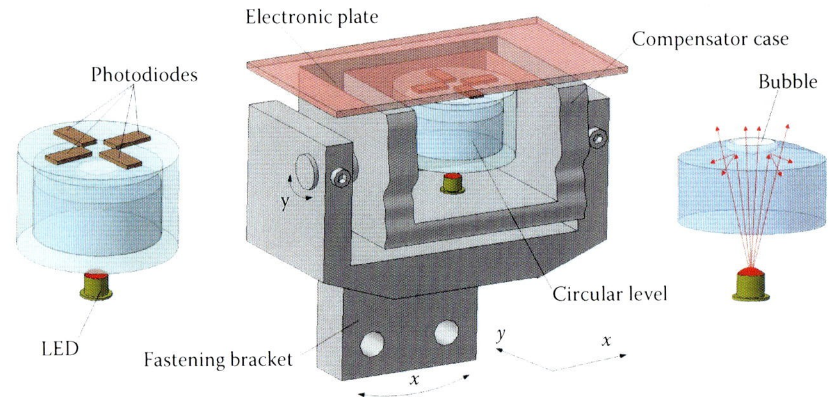

FIGURE 3.24 Dual-axes electronic compensator arrangement applied by Sokkia.

are set up on the electronic board together with amplifiers and a temperature sensor. The microprocessor applies these signals to calculate the bubble position. The bubble position information is available in either a graphic or digital form.

A dual-axial compensator is set up in the same location at a monoaxial unit, via the two fastening screws. It is adjusted along the x direction the same way as a monoaxial compensator. In order to adjust the compensator along the y direction, we turn the compensator case around its axis relative to the fastening bracket. Once the adjustment along the y direction is complete, the compensator is fixed with the help of the stopper screws.

Dual-axial compensator testing is very similar to monoaxial testing. Separate testing of both directions is carried out. The x direction testing is just like the monoaxial compensator testing. The y direction testing is also related, but the inclination angles are set in a different manner. First, we incline the theodolite along the x direction by rotating the foot screw and target the reticle to the index lines. Then we turn the theodolite at 90° and switch the display into electronic level mode. We can see the angular value of inclination along the y direction. We then switch the display into angle measuring mode and turn the theodolite 180°. Now we note the value of the inclination angle along the y direction. Then we correct the compensator along the y direction the way we did along the x direction. Of course, afterward we must finalize adjustment with the help of the compensator software and carry out the tests again.

A dual-axial compensator solution from Leica is shown in Figure 3.25.

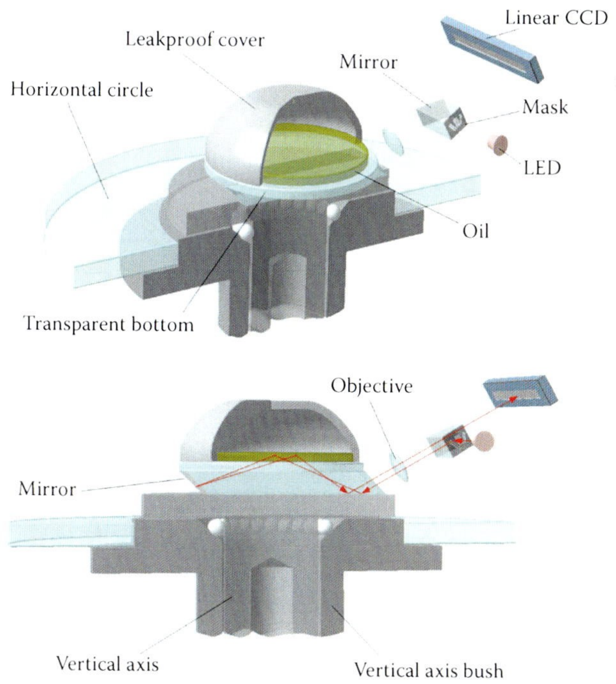

FIGURE 3.25 Dual-axes electronic compensator arrangement applied by Leica.

A leak proof vessel filled with silicone oil is used as a sensitive element in the compensator. The oil is used because oscillations subside quickly in it. The vessel has a transparent bottom. The upper surface of the oil has a mirror for the rays of light falling on the surface at an acute angle. LED emission is directed to an optical mask (Figure 3.26) that forms the image of orthogonal and inclined stripes. The image is turned with the mirror and goes through half of the objective. Then the image of stripes goes through the sensitive element and returns to the half of the objective that sends the image to the linear CCD.

There is an electronic signal at the CCD output (the lower part of Figure 3.26). The distance from the zero pixel to the central group of exposed pixels provides information about the *x* direction's incline. The interval between two groups of inclined lines provides the information about the incline of the *y* direction.

Other developers also use a vessel filled with silicone oil in their compensators. The solution brought forward by Trimble is shown in Figure 3.27. A narrow beam

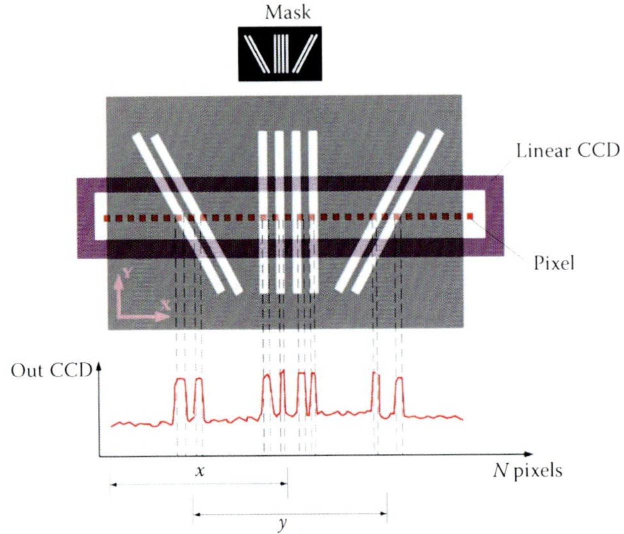

FIGURE 3.26 Readout principle of the compensator shown in Figure 3.25.

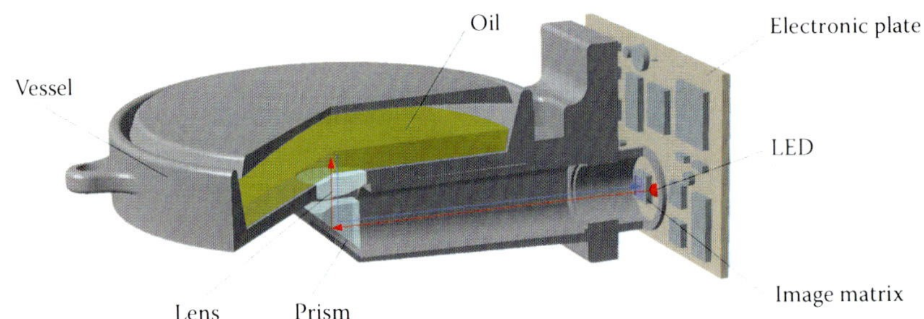

FIGURE 3.27 Dual-axes electronic compensator arrangement applied by Trimble.

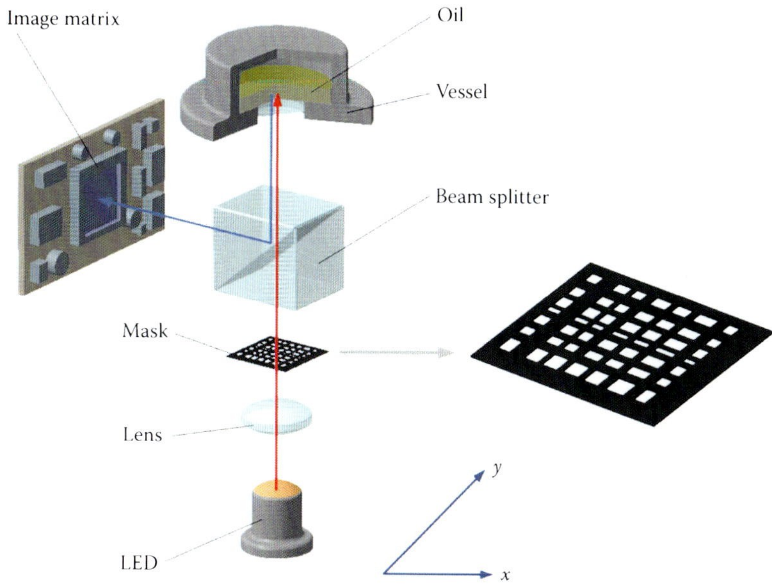

FIGURE 3.28 Dual-axes electronic compensator arrangement applied by Sokkia (new).

comes from an LED to the prism that rotates it to the vessel bottom. There is a lens window in the vessel bottom. The light beam is reflected from the oil's surface. Then the beam hits to the image matrix. A similar type is used in camcorders. There is a light spot on the sensitive area of the image matrix. A video signal comes from the matrix output to the image microprocessor that calculates the x and y coordinate of the light spot energetic center.

In the newest designs Sokkia applies the same structure for a dual-axes compensator (Figure 3.28). Their main difference consists in using a square mask that consists of two crossing orthogonal barcodes. The mask image moves around the sensitive area of the image matrix as a result of the oil surface incline change. The image microprocessor calculates image movement along both the x axis and the y axis. Typical programs are applied for barcode image processing.

Compensators that have a vessel filled with silicone oil and an image matrix (or the linear CCD) are more stable than those with a tubular (or circular) fluidal level. Also they have a wide operating range and better linearity. This is why they usually do not require mechanical adjusting. Periodically a program adjustment of the compensators is necessary with the purpose of reassigning their zero pixels.

3.5.3 Theodolite Telescope

Modern surveying instrument telescopes are often based on Kepler telescope principles. The story about its development and its optical scheme is in Chapter 2. In theodolites, 20 to 40 times magnification telescopes are used. This kind of magnification is necessary because the naked eye has angular resolution of about 30″, meanwhile required sighting accuracy in surveying is 2″ and above. We know that Kepler telescope magnification is described as

$$M = \frac{f_o}{f_e} \tag{3.4}$$

where f_o is the objective focal distance and f_e is the ocular focal distance.

There are some technological limits in the choice of ocular focal distance. It is difficult to make a short-focus ocular with acceptable geometrical distortions. That is why focus oculars of less than 10 mm are seldom applied in surveying instrument telescopes. If we insert this value in Equation 3.4 we will see that at 30 times telescope magnification its length is equal to 300 mm. Previous surveying instrument telescopes were quite large and long.

Nowadays surveying instrument objectives consist of two parts. There is a front objective and a focusing lens (see Figure 3.29).

Dual-lens optical systems have equivalent focal distance:

$$F = \frac{f_o f_F}{f_o + f_F - l} \tag{3.5}$$

where f_o is front objective focal distance, f_F is the focusing lens focal distance (if the lens is negative the minus sign "–" appears), and l is the distance between the front objective lens and the focusing lens.

When we analyze the formula we see that the equivalent focal distance F is longer than the focal distance for the front objective lens f_o. That means that in order to get the required telescope magnification, we must apply a shorter focal front objective lens and then add a negative lens that is set at the distance l following the front lens. Thus, the negative lens is applied for focusing. The total length of the telescope depends on the front objective lens's focal distance. Dual-component objective appliances allow us to shorten the telescope length by a rough factor of 2.

Modern telescope objectives may consist of three components. Telescopes of this kind are applied in surveying levels. Theodolites only have dual-component

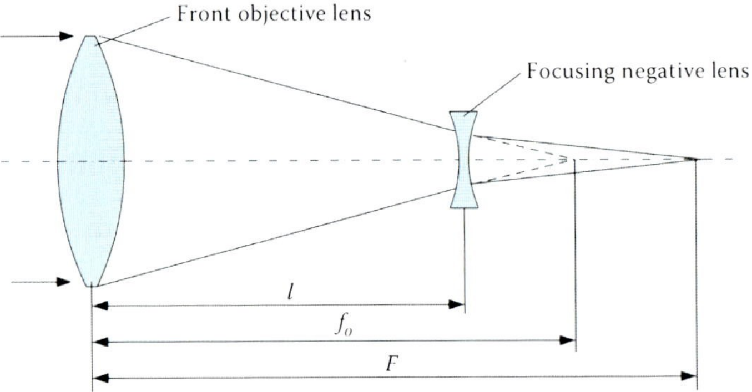

FIGURE 3.29 Dual-component objective.

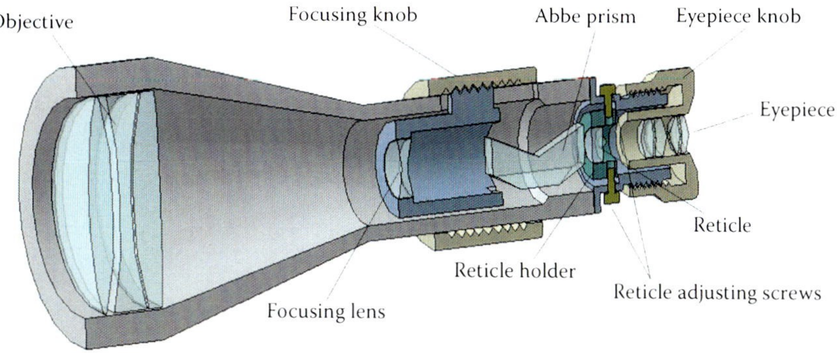

FIGURE 3.30 Theodolite telescope with Abbe prism.

objectives. Getting the direct image in theodolites is fulfilled the same way as with surveying levels. Optical schemes for converting reversed images into direct ones are described in Chapter 2. Abbe or Porro prisms are used for this goal (their full names are Abbe-Koefin or Porro-Abbe prisms).

Nowadays, in the majority of theodolites that have direct imaging, Abbe prism-type telescopes are applied (Figure 3.30).

This category of telescope consists of three main parts. These are the telescope's main body with the objective front lens, a focusing system, and the ocular element. The main body of the telescope also has axle journals, which are not present in the figure. The objective of theodolites usually has two or three lenses. Some of them consist of pairs of lenses that are joined together.

A theodolite focusing system consists of the focusing lens in a frame and a focusing knob. A cylindrical frame of the focusing lens has precise bearing slides allowing it to move along the optical axis of the telescope. The frame also has a cogged ledge connected with threading at the internal side of the focusing knob. As we rotate the focusing knob, the cogged ledge slides along the thread, causing the focusing lens to move.

A theodolite's ocular element consists of an ocular, a reticle, and an inverting prism. The ocular is placed into a frame, which can be moved within several millimeters along the telescope's optical axis by rotating the frame along the thread. Its movement is necessary for the reticle image individual focusing. The ocular consists of several lenses stuck together in pairs.

The reticle consists of two adhered round glass plates. The internal side of one of these plates has some crossed lines, whose thickness is 2 to 4 mkm. A two-plate solution is applied to protect the reticle from dust. The reticle is put into a frame, which can move along two directions by means of four adjusting screws. The direction of movement is perpendicular to the telescope's optical axis.

Reticle adjusting units are of the pushing or pulling type. The pulling type is more popular now because in the pushing type reticles can be destroyed with excessive tightening of the adjusting screws.

The inverting prism is associated with the ocular part in that it is usually set on top of it. As mentioned earlier, besides the Abbe prism, an inverting Porro prism may

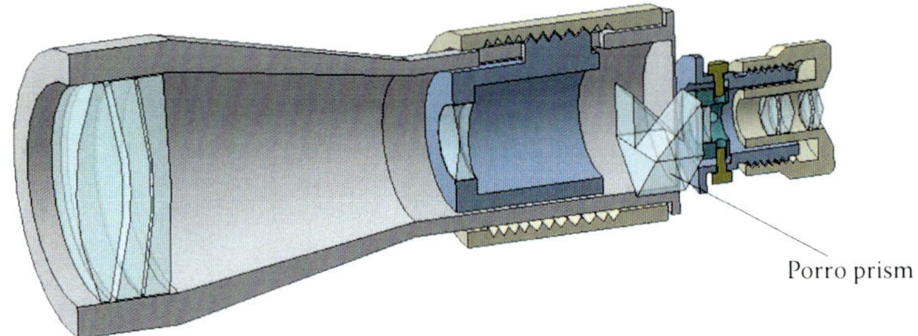

FIGURE 3.31 Theodolite telescope with Porro prism.

also be used in theodolites (Figure 3.31). The Porro prism is quite often applied in total stations, however only Nikon uses it in theodolites. Telescopes fitted out with a Porro prism are a bit shorter than those equipped with Abbe prisms. The Porro prism solution brings about the removal of the ocular axis relative to the objective axis.

Laser theodolites allow target visualization while fulfilling the layout. A direct image telescope with a built-in laser module is the primary component of a contemporary laser-type theodolite (Figure 3.32). Laser and sighting channels are separated by a splitting prism. This prism consists of two attached one-half fractions of a glass cube. The internal side of one of these fractions has a monochromatic mirror covering. It reflects only laser spectrum beams, otherwise it is transparent for optical beams in the visible range. The splitting prism is located between the focusing lens and Porro (or Abbe) prism. That is why image and laser spot focusing occur simultaneously.

The source of light from the laser module is placed at the same distance from the objective as the reticle. Therefore, at the moment the telescope is pointed to the target, it is illuminated by the focused laser spot. Unfortunately, the part of the laser light reflected from the objective lenses goes through the splitting prism and illuminates the target with a red aureole. In order to eliminate this effect, we suggest setting a protective red spectral filter onto the ocular at the moment the laser is switched on. The removable spectral filter is available in the complete theodolite set.

The most well-known laser theodolites are listed in Table 3.4.

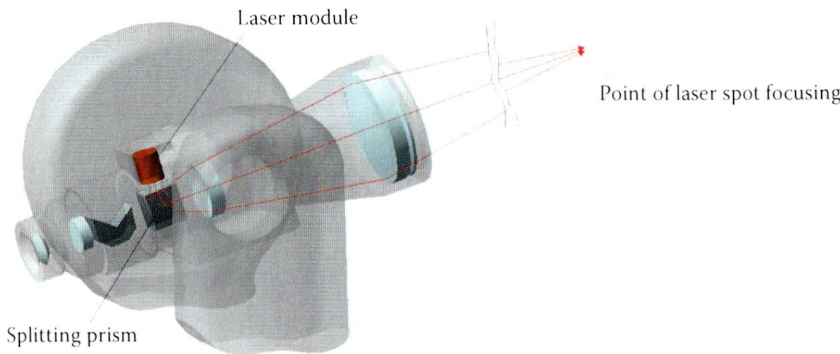

FIGURE 3.32 Theodolite telescope with a laser pointer.

TABLE 3.4
Up-to-Date Electronic Laser Theodolites

Model	Class of Accuracy	Value of Accuracy (n'')	Manufacturer Website
DJD2-JEL	High	2	http://www.boif.com
LP212	High	2	http://www.foif.com
ET02L	High	2	http://www.southinstrument.com
LDT520	Moderate	5	http://www.sokkia.com
DT-205L	Moderate	5	http://www.topcon.com
LP215	Moderate	5	http://www.foif.com
ET05L	Moderate	5	http://www.southinstrument.com
DT-207L	Moderate	7	http://www.topcon.com
DL209L	Moderate	9	http://www.topcon.com

3.5.4 Theodolite Targeting Units

The integral part of any theodolite is its targeting unit. A separated targeting and clamping screws solution was applied in the earliest theodolites and continued to be used for a long time. Such a solution is also applied in many contemporary optical theodolites. Nowadays, all low-precision optical theodolites have such a targeting system (Figure 3.33).

The most recently developed theodolites have more ergonomic coaxial targeting solutions. This is mainly in reference to high- and moderate-precision theodolites. Even some of these theodolites have a separated screws system. These are usually theodolites that are produced under license. Up-to-date electronic theodolites are released with coaxial screws only. Their horizontal targeting solutions have the same operating principle. The coaxial screw arrangement as shown in Figure 3.34 and is typical for Sokkia and Pentax theodolites. Topcon and Nikon instruments also have

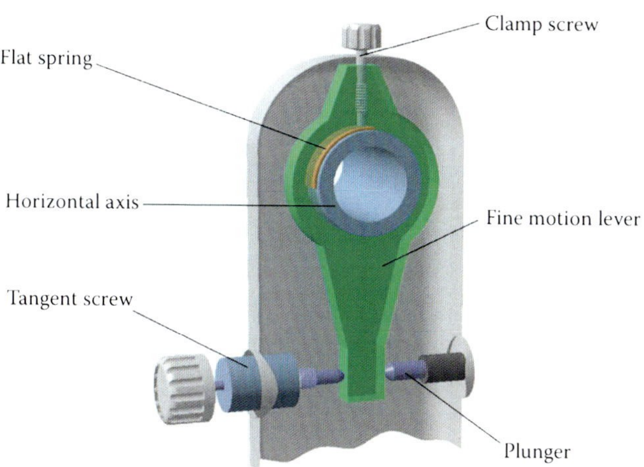

FIGURE 3.33 Targeting unit provided with separated screws.

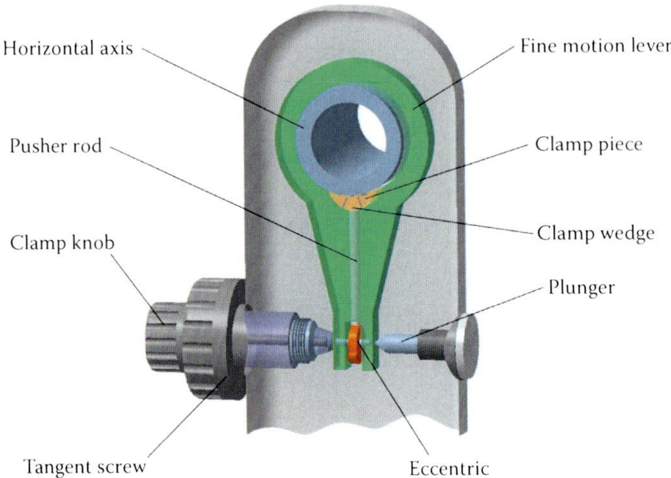

FIGURE 3.34 Targeting unit provided with coaxial screws.

similar coaxial screws. Topcon and Nikon solutions have a thin targeting screw that is set in the inside of the clamping screw.

Leica theodolites have so-called infinite screws (Figure 3.35). Here you can see the targeting unit arrangement for the theodolite horizontal axis. A worm gear is used for accurate sighting. In order to establish preliminary aiming, we should make some effort to turn the theodolite getting over the wavy spring braking force. A good aspect of the solution is faster targeting. As far as the efforts made to turn the theodolite are considerable, the foot screws and the tripod stability must meet high requirements (Figure 3.36).

If the error quantity from 10″ to 60″ errors occur while measuring horizontal angles, we should check the leveling screws and the tripod. As needed, we should adjust them. Surely they must be checked while using any theodolite; however, it is the theodolites with infinite targeting screws that are especially sensitive to these errors.

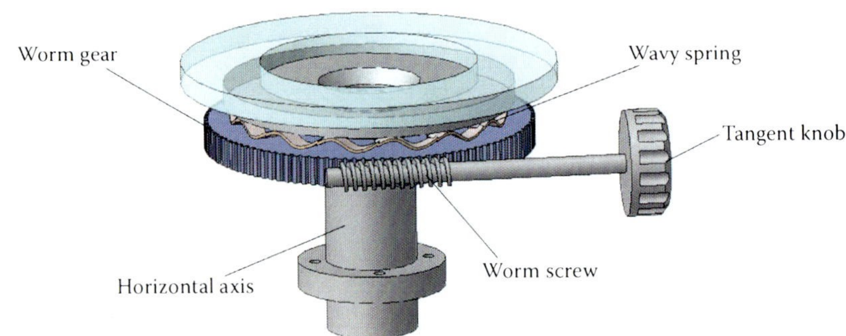

FIGURE 3.35 Infinite tangent screw.

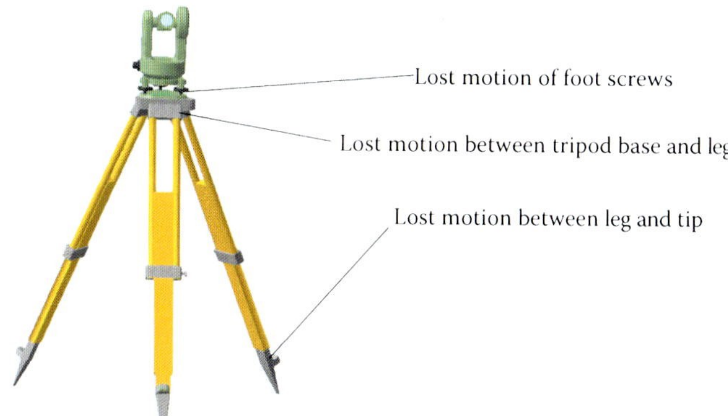

FIGURE 3.36 Reasons for horizontal angle errors.

3.5.5 Theodolite Plummets

In order to set the theodolite exactly over the point of reference, in modern times we apply built-in optical and laser plummets. The optical plummet is the Kepler telescope provided with an internal focusing lens (Figure 3.37). The direct image is achieved by the right-angled roof prism that also directs the plummet optical axis vertical down. The bush of the theodolite's vertical axis is hollow. The plummet optical system's magnification is usually about three times. The telescope reticle is superposed with the theodolite vertical axis with four adjusting screws.

The superposed accuracy of the theodolite's vertical axis, with the plummet axis, is assessed the following way (see Figure 3.38). We place the tripod with the

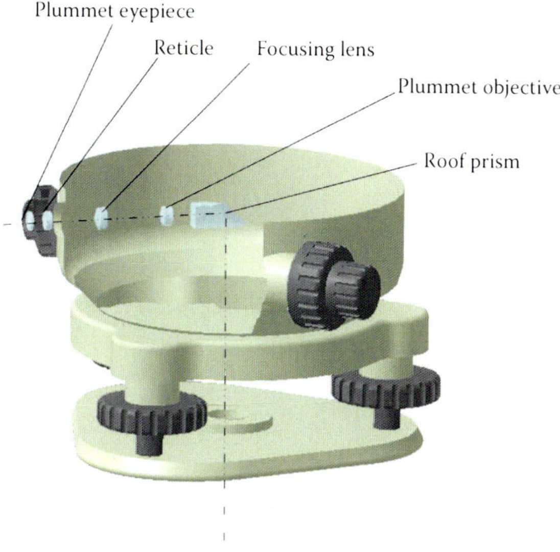

FIGURE 3.37 Theodolite optical plummet.

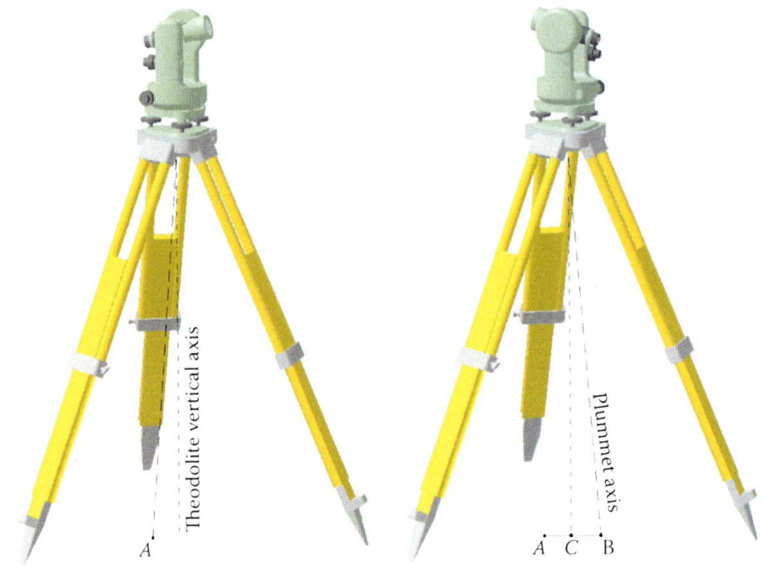

FIGURE 3.38 Checking up the theodolite plummet.

theodolite on a plane surface and mark point *A* by means of the plummet. At this time we do not pay attention to the position of the bubbles. Then we turn the theodolite to 180° and mark point *B*. If we split the distance *AB*, we get point *C*, which is found on the theodolite's vertical axis. Then we should superpose the reticle with point *C* by adjusting the reticle screws. Again, we now rotate the instrument to 180°

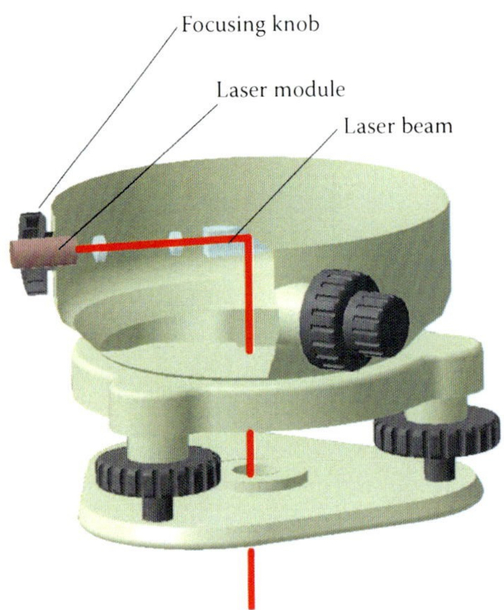

FIGURE 3.39 Theodolite laser plummet.

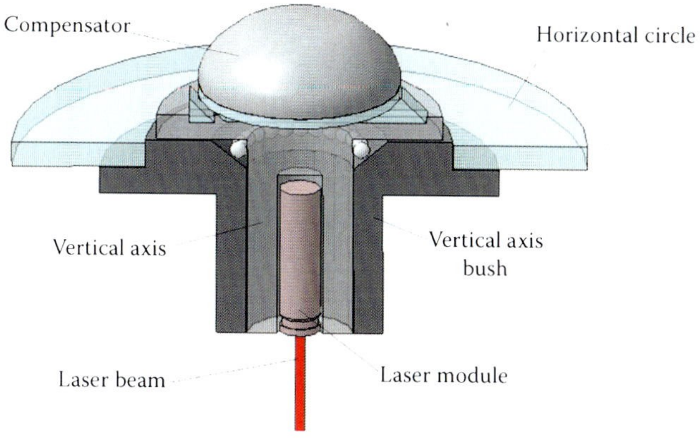

FIGURE 3.40 Laser theodolite plummet applied by Leica.

and check whether the reticle is removed from point *C*. If so, we should complete the adjusting steps once more.

Optical plummets of this kind can be easily converted for laser use (Figure 3.39). Manufacturers install a laser module instead of the reticle and the ocular. Checking and adjustment are carried out in the same manner that the optical plummet is tested. In this case, the adjusting screws remove the laser module case and not the reticle.

These days, optical plummets are mainly built into the theodolites and seldom are on the tribrach. Optical plummets that are built into tribrachs are more typical for low precision theodolites. It is difficult to adjust a plummet that has been built into the tribrach. It is often advised to adjust the plummet by placing the theodolite sideways onto a table edge, and then turning the tribrach itself to 180°. The points are marked on a paperboard set up at 1.5 m distance. We do not advise this sort of adjustment for a theodolite, as it is difficult to fasten it properly at the edge of the table. We risk dropping the instrument. It would be better to use some other accessories such as a prism reflector holder or an angle measuring mark.

We can also adjust this plummet by the means of tribrach removal at 120°. We put the theodolite on the tripod and then properly set the horizontal aspect. Then we outline the contour of the tribrach on the tripod base. Now we mark a point according to the plummet reticle on the paperboard, which is placed under the tripod. Now we slightly loosen the fastening screw and turn the tribrach to 120°. Then we accurately superpose the tribrach with its contour on the tripod base. Again, we set the horizontal aspect of the theodolite and mark the second point on the paperboard. We use this same procedure to get the third point. Afterward, we find the triangle's center and superpose the reticle with it by fine-tuning the adjusting screws.

Laser plummets built into a theodolite's vertical axis are considered to be the most current and accurate (Figure 3.40). Here we see that it is very well protected and does not demand adjustment. Coincidence between the theodolite's vertical axis and the laser beam are guaranteed by the manufacturer.

BIBLIOGRAPHY

Anderegg, J. S. 1966. Shaft encoders. US Patent 3,244,895 filed July 26, 2962, and issued April 5, 1966.

Glimm, A. 2006. Inclination detection method and apparatus. US Patent 2006/0170908 A1 filed January 10, 2005, and issued August 3, 2006.

Gohdo, E., T. Maezawa, and M. Saito. 1999. Laser theodolite. US Patent 5,905,592 filed August 28, 1997, and issued May 18, 1999.

Hori, N. and T. Yokoura. 1986. Tilt angle detection device. US Patent 4,628,612 filed October 1, 1985, and issued December 16, 1986.

Imaizumi, Y. 1994. Rotation angle measuring apparatus. US Patent 5,301,434 filed December 17, 1992, and issued April 12, 1994.

Ishikawa, Y. and M. Tanaka. 1984. Optical system of a theodolite. US Patent 4,445,777 filed January 17, 1983, and issued May 1, 1984.

Kumagai, K. 2004. Absolute encoder. US Patent 6,677,863 B2 filed April 3, 2002, and issued January 13, 2004.

Larsen, H. R. 1944. Theodolite. US Patent 2,363,877 filed February 11, 1943, and issued November 28, 1944.

Leitz, A. 1902. Transit. US Patent 715,823 filed May 21, 1901, and issued December 16, 1902.

Ley, C. H. 1915. Adjusting device for theodolite. US Patent 1,145,075 filed March 16, 1915, and issued July 6, 1915.

Lippuner, H. 2006. Tilt sensor. US Patent 2006/0005407 A1 filed July 12, 2005, and issued January 12, 2006.

Matsumoto, T. and K. Ohno. 1996. Absolute encoder having absolute pattern and incremental pattern graduations with phase control. US Patent 5,563,408 filed October 27, 1994, and issued October 8, 1996.

Ohishi, M. 2001. Tilt detecting device. US Patent 6,248,989 B1 filed April 28, 1998, and issued June 19, 2001.

Ohtomo, F. and K. Kimura. 1984. Apparatus for measuring length or angle. US Patent 4,484,816 filed July 20, 1983, and issued November 27, 1984.

Peter, J. and E. Kooi. 1980. Theodolite tangent screw system. US Patent 4,202,110 filed May 18, 1979, and issued May 13, 1980.

Sawaguchi, S. 2003. Laser centering device. US Patent 2003/0177652 A1 filed January 22, 2003, and issued September 25, 2003.

Shimura, K. 1992. Inclination angle detector. US Patent 5,101,570 filed July 14, 1989, and issued April 7, 1992.

Wild, H. Angle-measuring instrument. US Patent 2,221,317 filed January 26, 1938, and issued November 12, 1940.

Wingate, S. A. Photoelectric shaft angle encoder. US Patent 3,187,187 filed January 24, 1962, and issued June 1, 1965.

4 Electronic Distance Measurer (EDM)

Until the middle of the last century, accurate surveying of linear measurements was very expensive and complicated to carry out. This is why geodetic network structures comprising a minimal amount of measuring lines were chosen. In some cases, geodetic network sides were not available for measuring using mechanical measuring instruments due to natural barriers (ravines, rivers, etc.). *Inaccessible distance measurements* were a serious issue.

Indirect optical measuring techniques were not developed until the beginning of the last century. Their application was initially limited due to insufficient accuracy of their measuring ability. The best instrument of this kind was the Redta 002, which had maximal accuracy of 1:7000 and several hundred meters of ranging.

The introduction of electronic distance measurers (EDMs) into surveying practice became a real technical revolution. At first these instruments were bulky and expensive. The first surveying light-measurer devices looked like a field lab weighing up to 200 kg and having their own power plant. Measurements were carried out at nighttime. It took about half an hour to measure one line. Then again, they measured distances up to 50 km with an accuracy of within 1:850,000! Powerful mercurial lamps, and later gas lasers, were used as the sources of light. Photoelectronic multipliers, and rarely just simply the human eye, had been used as photoreceivers for quiet some time.

In 1951 Swedish physicist E. Berdstrand was the first to develop an EDM. Jointly with manufacturer AGA, he produced several types of EDMs under the brand Geodimeter NASM (see Figure 4.1). They were suitable for surveying. Those instruments were able to measure distances of tens of kilometers with a relative accuracy of about 1:300,000.

Early EDMs could only provide the required precision and distance range through the use of optical reflectors. That was because of imperfect light resources (various kinds of lamps) and the low sensitivity of the optical detectors. Those light sources were powerful, but they had large luminescence bodies and low operation speed. That is why they were only applied with external high-speed light modulators. In order to get an acceptable divergence of light rays, it was necessary to use bulky long-focused optical systems.

Several kinds of reflectors were used together in the first geodimeters (Figure 4.2). Corner cube prism reflectors turned out to be the most practical (Figure 4.3). They are small in size and have an acceptable reflected beam divergence of several angular seconds. Besides it can be applicable even if it is turned within 30°. Other types of reflectors can only be directed to the EDM within several angular minutes. Therefore, because of their effectiveness corner cube prism reflectors are the type of units that we use in modern times.

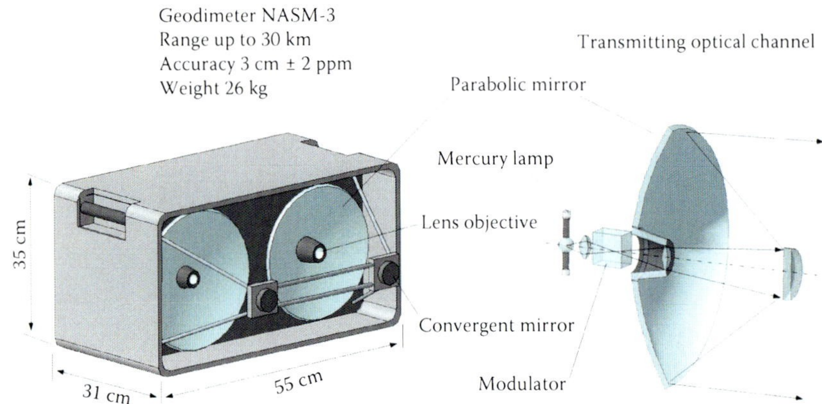

FIGURE 4.1 One of the first EDMs.

FIGURE 4.2 EDM reflectors.

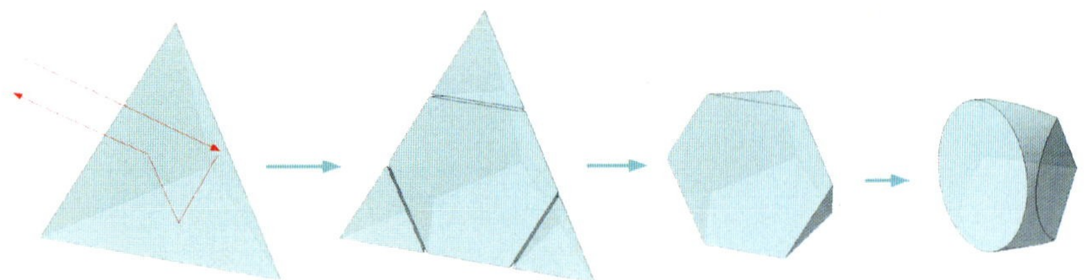

FIGURE 4.3 Corner cube prism reflector.

Application of small semiconductor lasers became an important step in the development of EDMs. This started in the 1970s. At that time only infrared semiconductor lasers were used. Since the 1980s, photoelectronic multipliers have been excluded by avalanche photodiodes (APDs). About the same time an EDM was joined together with a theodolite. A new surveying instrument was born: the prototype for an electronic total station (Figure 4.4).

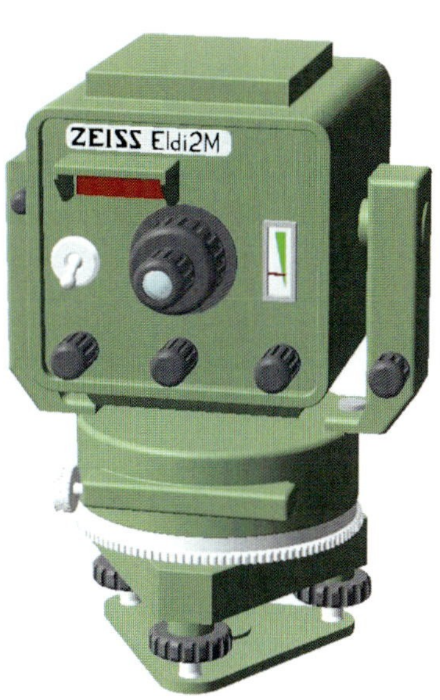

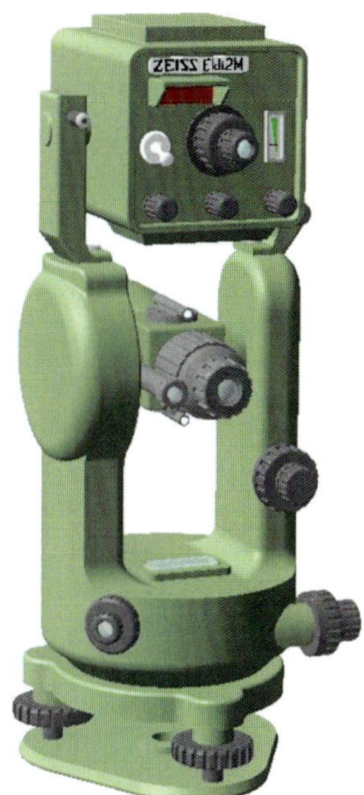

FIGURE 4.4 The EDM on the tribrach and on a theodolite.

Use of a red semiconductor laser as the light source is more than likely the last important step in the development of EDMs. Nowadays such EDMs are often used as a component in total stations and are able to work with or without prisms.

Since the 1990s portable handheld EDMs started to appear. Red lasers are applied in these units. EDMs of this type are able to measure lines at direct reflection mode up to 200 m with 1–5 mm accuracy. Use of a prism is not provided in them.

4.1 PHYSICAL FUNDAMENTALS OF SURVEYING ELECTRONIC DISTANCE MEASURERS (EDMs)

In modern surveying practices about 95% of linear measurements are fulfilled applying EDMs. The physical principles behind the measurement of distances using EDMs are simple. We send a ray of light to an object and part of the light energy bounces back to us. Then we measure the time, t, for the light sent to the object and backward. The time value is then inserted into a basic formula, and the unknown distance referred to as S is calculated:

$$S = \frac{c \cdot t}{2n} \tag{4.1}$$

where c is the light speed in the vacuum (299,792,458 m/s); and n is the index of air refraction, which depends on light wavelength, atmospheric temperature, pressure, and humidity. On average the n index is equal to approximately 1.0003.

The n index is calculated in two steps. First, we calculate the index of the wavelength λ for the appointed source of light at standard meteorological conditions, for example, absolutely dry air at the temperature of 0°C and atmospheric pressure of 1013.25 mbar. This step is carried out in the EDM design stage when the source of light is being chosen.

For the second step, the obtained value is substituted into the Barrel and Sears formula for further calculation of the true atmosphere refraction index.

Instead of the n characteristic for refraction, we often apply a more convenient value $N = (n - 1) \cdot 10^6$, which is referred to number refraction, or atmospheric correction. Here we will give two examples of working formulas that are applied in modern total stations computers to calculate EDM atmospheric correction data.

For example, in the total station Leica TPS1200 ($\lambda = 658$ nm) we apply the formula comprising of a complete set of atmospheric parameters, air humidity included:

$$N = 286.34 - \left[\frac{0.29525 \cdot p}{1 + \alpha \cdot t_c} - \frac{4.126 \cdot 10^{-4}}{1 + \alpha \cdot t_c} \cdot h \cdot 10^{x} \right] \quad (4.2)$$

where N is the refraction number (ppm), p is the air pressure (mbar), t_c is the air temperature (°C), and h is the relative humidity (%). Also

$$\alpha = \frac{1}{273.15}$$

$$\chi = \left(\frac{7.5 \cdot t_c}{237.3 + t_c} \right) + 0.7857$$

For precision distance measurements the atmospheric correction should be determined with an accuracy of 1 ppm. It requires an air temperature accuracy within 1°C and an air pressure within 3 mbar. The air humidity is taken into consideration at combinations of high temperature (more than 30°C) with high humidity (more than 80%). In the case of routine measurement accuracy within moderate atmospheric conditions, we can ignore air humidity factors.

In our second case, regarding total station Topcon GPT-3100N ($\lambda = 870$ nm), the formula is deduced with no account taken of air humidity:

$$N = 279.85 - \left[\frac{79.585 \cdot p}{273.15 + t_c} \right] \quad (4.3)$$

where N is the refraction number (ppm), p is the air pressure (mbar), and t_c is the air temperature (°C).

Obtaining the measurement of t is a very difficult technical problem. Light at the speed of c/n passes twice the distance of 1 km in a period 6.6 microseconds. That means if we need to measure a distance within 1 mm accuracy, we must measure the time interval of t with a $6.61 \cdot 10^{-12}$ accuracy!

EDM developers have another important problem to solve, which the main formula (Equation 4.1) does not describe. It is necessary to get a required intensity signal reflected from the object. In order to measure the time interval t we need to get the signal that several times exceeds noise. The problem is solved by choosing a source of light, a photoreceiver, and an optical system.

Contemporary surveying EDMs employ semiconductor-based lasers as their light source (see Section 4.3.3). These lasers have a high coefficient of performance, narrow-band spectrum emissions, and project luminescent body of only a few microns in size. Two types of semiconductor lasers are primarily used. The first type is continuously emitting laser whose optical range is 630–670 nm. Its maximum power emission is limited to 5 mW because of medical reasons. The second type is an impulse infrared laser (820 nm) whose impulse power is 3–5 W. Since the EDMs laser pulse spacing is significantly larger than the pulse duration, an averaged emission does not exceed eye safety limits.

The main element of an EDM is its optical detector. Only APDs are applied in current surveying EDMs. The principles behind them are similar to ordinary semiconductor photodiodes. The concepts are described in detail in Section 4.3.2. An electrical signal appears when lighting the semiconductor p-n junction. Nevertheless the APDs are essentially different from ordinary photodiodes. High reverse bias voltage is applied to their p-n junctions (from 100 to 400 V depending on the APD make). Such voltage induces avalanche electrical p-n junction breakdown. The current limiting resistor of the APD supply circuit does not allow the breakdown and destruction of the p-n junction. We should work in the operation mode when the bias voltage is a little less than the avalanche breakdown voltage. This operating mode is called preavalanche. An APD operating in this mode has a maximal sensitivity and speed. The typical sensitivity of an APD is about 30 A/W for visible red light and about 50 A/W for infrared light. In addition, APDs have very low level of intrinsic noise (dark current is about 10^{-10} A) and high coefficient of efficiency (75%).

As stated earlier, an effective signal must exceed noise by several times for exact time interval measurements. We would seek 10 times the proportion between signal and noise for our calculations to be considered reliable. That means we should provide such a signal level of light entering the APD to produce a current equal to 10^{-9} A. It is about $7 \cdot 10^{-11}$ W for red light, and $3 \cdot 10^{-11}$ W for infrared light. The optical signal level entering the optical detector depends on the light source power, reflecting capacities of the object, the atmospheric transparency, and the EDM optical scheme's effectiveness.

At the present moment, surveying EDMs equipped with prism reflectors have not produced any more than as a separate unit. Modern EDMs as a part of a total station are able to work with a prism reflector and also in a direct reflection mode. Red and infrared lasers are used in these EDMs.

There is a class of handheld EDMs that measure distances in the direct reflection mode only. These models mainly use red visual light lasers. The maximal range of surveying EDMs is usually from 100 to 500 m in direct reflection mode, and up to 10 km when using a prism reflector. There are some EDMs that can operate in direct reflection mode at a distance of 2 km!

Real-world objects generally have diffusive reflecting surfaces and have poor mirror characteristics. When the objects are illuminated up a laser, they can be considered as a diffusive light source. According to Lambert's law, the diffusive source luminance is the same in all directions. Also Lambert suggested a whiteness coefficient that indicates how much light was reflected by the surface. The rest of the light is absorbed. The albedo (the Latin word meaning "whiteness") coefficient is about 1 for light surfaces (for snow it is 0.8; for white paper it is 0.7), and approaches 0 for dark surfaces (for ground it is 0.15; black velvet is 0.05). Since some objects may have reflective features, the luminous power going along from the reflecting surface is two to four times more than in case of an ideal diffusive surface. For this reason a direct reflection EDM range depends on the surface inclination relative to the laser beam.

According to Lambert's law, the receiving objective is able to collect only part of the luminous power, which is limited by the solid angle ω (Figure 4.5).

Taking into account the albedo coefficient α and atmospheric transmittance τ, the luminous power F_{ob} collected by the receiving objective is

$$F_{ob} = \frac{\alpha \tau^{s} F_t \pi D^2}{4S^2} \tag{4.4}$$

where F_t is the luminous power entering the target, D is the receiving objective diameter, S is the distance to the target, π equals 3.14, and τ^s is the atmospheric transmittance (in this case the distance to the target S measured in kilometers).

The atmospheric transmittance depends on the atmosphere's condition (purity of the air) and also the wavelength of the light used. The average atmospheric condition visual range of 20 km is often taken for calculation. Relative spectral transparency of visual and nearest infrared are equal to 0.8 and 0.9, respectively.

We can calculate the luminous power at the output of the transmitting objective that reaches the target. Luminous power that reaches the target after the deduction of light dispersed and adsorbed by the atmosphere looks like

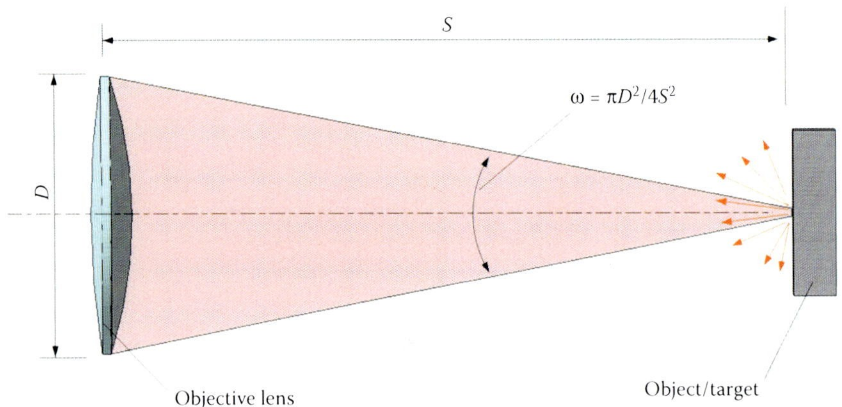

FIGURE 4.5 Luminous power collected by the objective lens.

$$F_t = \tau^s F_{out} \tag{4.5}$$

where F_{out} is the luminous power at the output of the transmitting objective.

The reflected light collected by the receiving objective does not fully reach the optical detector because of losses at the receiving end of the optical system and at the narrow-band light filter. This light filter is usually placed directly in front of the optical detector and is intended for the removal of background lighting influences. Total losses of light in the receiving channel are usually half of the amount collected by the receiving objective. Considering this, the luminous power F_{in} reaching the optical detector is

$$F_{in} = 0.5F_{ob} = \frac{0.5\alpha\tau^{2s} F_{out} \pi D^2}{4S^2} \tag{4.6}$$

Applying this formula we can get a valuation formula for an EDM's maximal operating range in direct reflection mode:

$$S = 0.62\tau^s D\sqrt{\frac{F_{out}}{F_{in}}} \tag{4.7}$$

Equation 4.7 has S in both parts of the equalities, that means according to mathematical conceptions they are transcendental. In practice these evaluations are solved with the help of iterative methods. For the first iterative step in our calculations we can apply a value of atmospheric transmittance as τ^s from Table 4.1. Then we insert a calculated preliminary value of S into Equation 4.7 and get a final value of S. Two iterative steps are enough to get an estimated value of the EDM's maximal operating range.

Equation 4.5 can be applied only to determine a reflectorless EDM's maximal operational range estimation, as the laser beams influences are negligible at short distances. Equation 4.5 contains no component considering divergence of an emitted laser beam.

Later we will demonstrate that divergence should be taken into consideration only while the EDM is operating in prism mode and when large distances are measured. In order to understand the nature of divergence we refer to Figure 4.6.

TABLE 4.1
Atmospheric Transmittance Values for Red and Infrared Lights

Source of Light	Distance to the Target (km)					
	0.1	**0.2**	**0.5**	**1.0**	**2.0**	**5.0**
Red	0.98		0.89	0.80	0.64	0.33
Infrared	0.99	0.98	0.95	0.90	0.81	0.59

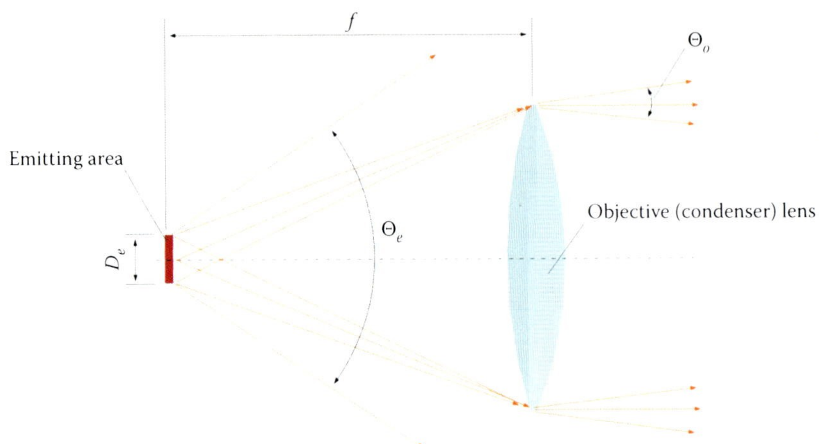

FIGURE 4.6 Explanation of laser beam divergence.

Laser-emitting areas in semiconductors are a luminescent stripe about 1.5 × 6 μm in size. The emissions divergence maximum angle Θ_e is within 6° to 60° for different planes. As a rule, the condenser lens does not collect all laser emitted light but only a part of it within the solid angle of 30° to 45°. If the laser's luminescent area is in the condenser lens's focus, the beams shed from the center area become parallel at the condenser lens output. The beams shed by the edges of the luminescent area diverged at an angle:

$$\Theta_o = \frac{D_e}{f} \tag{4.8}$$

where D_e is the luminescent area size collected by the condenser and f is the condenser lens focal distance.

EDMs with a red laser have condenser lenses whose diameters are of 4–7 mm. Their focuses are 5–10 mm. In this case the divergence of laser pencil of rays is within 1/3000 and 1/1000 (about 0.3–1.0 mrad).

The best models of EDMs are provided with a visible beam to ensure a maximal distance up to 400 m and minimal laser beam divergence. The laser spot size on the object is modified from 20 mm (at the distance of 50 m) to 200 mm (at the distance of 400 m). Other EDMs ensure maximal distance range up to 200 m. Their laser spot size is no more than 200 mm, whereas their beam divergence is 1 mrad. A laser spot of this size is usually less than the target size and the laser beam divergence is not taken into account. That's why Equation 4.5 is correct in this case.

If a prism reflector is used we should take into consideration laser beam divergence. The formula changes slightly in this case. In order to deduce the formula, we use Figure 4.7.

In Figure 4.7 there is a coaxial optical scheme of an EDM. This is the arrangement primarily used in total stations. Transmitting and receiving optical channels

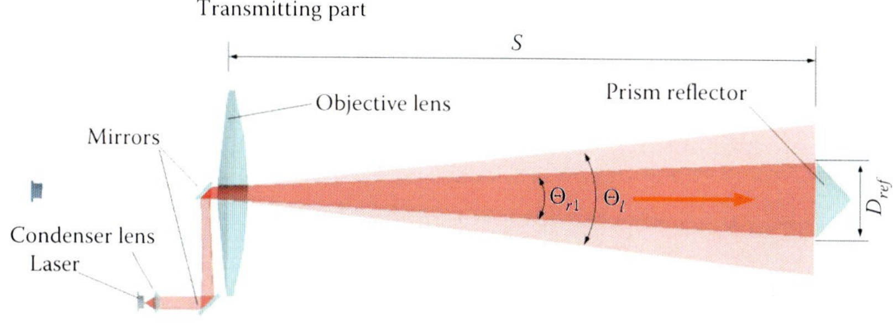

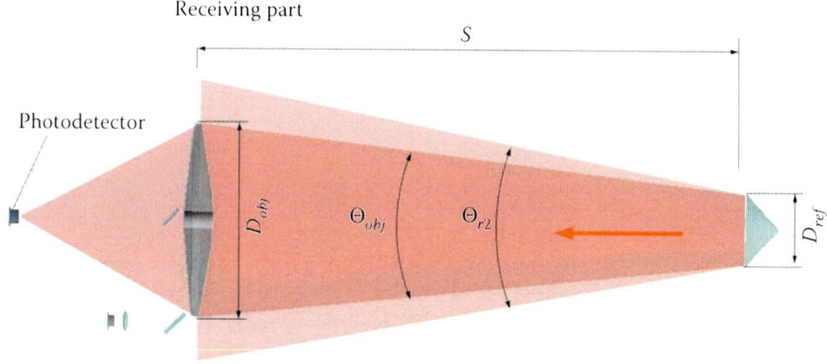

FIGURE 4.7 EDM optical parts.

are superposed along the same axis by the means of mirror systems. The beam path in the transmitting channel is demonstrated in the upper part of Figure 4.7. The spot of the laser beam whose divergence Θ_l is equal to 1 mrad at the distance S of over 65 m is more than the typical prism reflector diameter D_{ref} (65 mm). In this case, only a part of optical energy reaches the reflector prism. Taking into consideration Equation 4.3, we can express the interpretation like this:

$$F_r = \frac{\tau^s F_{out}\omega_{r1}}{\omega_l} \tag{4.9}$$

where F_r is the luminous power reaching the reflector, ω_l is the solid angle where the laser luminous power is emitted, and ω_{r1} is the solid angle where the part of the laser emission reaching the reflector is present.

For a divergence angle of $\Theta < 60°$ the correlation of solid and plain angles are expressed exactly like this:

$$\omega = \frac{\Theta^2\pi}{4} \tag{4.10}$$

Taking this into account we can express Equation 4.9 as

$$F_r = \frac{\tau^s F_{out} \Theta_{r1}^2}{\Theta_l^2} = \frac{\tau^s F_{out} D_{ref}^2}{S^2 \Theta_l^2} \tag{4.11}$$

Next we look at the path of the receiving optical element in an EDM. It is demonstrated at the bottom of Figure 4.8. We shall consider the prism reflector as a source of light. The prism reflector's transmission factor is 0.8. Therefore, the luminous power sent backward to the EDM is estimated as 0.8 F_r.

Divergence of the light beams emitted by the prism reflector is Θ_{r2}. A prism reflector has little influence on divergence of the pencil rays entering it, because the prism reflector's proper divergence is less than 20″, whereas the entering light divergence is 1 mrad (about 206″). So we can allow $\Theta_{r2} = \Theta_l$ in our calculation. The receiving objective collects only part of the light power sent back by the prism reflector. We mark the luminous power collected by the receiving objective as F_{ob}.

$$F_{ob} = \frac{\tau^s 0.8\ F_r \Theta_{obj}^2}{\Theta_l^2} = \frac{\tau^{2s} 0.8\ F_{out} D_{ref}^2 D_{obj}^2}{S^4 \Theta_l^4} \tag{4.12}$$

Taking into consideration losses of light in the receiving optical channel, the luminous power reaching the photodetector is expressed as

$$F_{in} = 0.5\ F_{ob} = \frac{n\tau^{2s} 0.4\ F_{out} D_{ref}^2 D_{obj}^2}{S^4 \Theta_l^4} \tag{4.13}$$

where n is the number of prisms in the reflector.

Thus, we have an estimating formula for the maximum range of the EDM operating in prism mode:

$$S = 0.8 / \Theta_l \sqrt[4]{\frac{nF_{out} \tau^{2s} D_{ref}^2 D_{obj}^2}{F_{in}}} \tag{4.14}$$

In the case of applying Equation 4.5 we can choose values from Table 4.1 for calculations of τ^s.

Now we have formulas to describe an EDM's maximum range in direct reflection mode (Equation 4.7) and in prism mode (Equation 4.14).

These formulas are not as strict as the Gauss distribution of power in light, so the pencil of ray is not considered. Nevertheless these formulas are applicable for practical use.

We move on to verification. Let us estimate the EDM distance range in reflectorless mode. For this we set the following parameters for our EDM:

- The red laser light power F_{out} at the optical system output is 1 mW (10^{-3} W).
- The receiving objective lens diameter D_{obj} is equal to 50 mm ($5 \cdot 10^{-2}$ m).
- The optical detector sensitivity F_{in} is equal to $3 \cdot 10^{-11}$ W for red light.
- Red light transmission of the atmosphere at the distance of 200 m τ^s is equal to 0.96.

Now we insert these values into Equation 4.7:

$$S = 0.62\tau^s D_{obj}\sqrt{\frac{\alpha F_{out}}{F_{in}}} = 0.62 \cdot 0.89 \cdot 5 \cdot 10^{-2} m\sqrt{\frac{\alpha \cdot 10^{-3}\,W}{3 \cdot 10^{-11}\,W}} = 170\ m\sqrt{\alpha}$$

The albedo coefficient α depends strongly on reflecting surface characteristics. Also we should consider mirror characteristics of surfaces. This was discussed earlier. We describe them as "mirror coefficients." For example, we calculate S for thee surfaces:

- Black paper: $\alpha = 0.05$; mirror coefficient 2
- White paper: $\alpha = 0.75$; mirror coefficient 2
- Special reflecting plate: $\alpha = 0.95$; mirror coefficient 4

Maximal operating range S for these tree surfaces is, respectively, equal to 53, 208, and 331 m. It corresponds to practice and Equation 4.7. As we can see, the calculation was made for an EDM with a 1 mW red laser. This sort of power limited laser is set in a portable laser rangefinder and also in older EDMs on total stations. The laser power of current EDMs on newer total stations has increased to 5 mW. Now we will make a new evaluation of Equation 4.7. For this we set new parameters for our EDM:

- The red laser light power F_{out} at the optical system output is 1 mW (10^{-3} W).
- The red light transmission of the atmosphere at the distance of 500 m τ^s is equal to 0.89.

The other parameters remain the same. New values for maximal operating range S are for black paper 110 m, for white paper 440 m, and for a special reflecting plate 670 m.

As said before, there are direct reflection EDMs whose maximal operating range is up to 2000 m. These pulse EDMs have an electronic optical distance measurement system based on the waveform digitizing (WFD) technology. This technology is mentioned in Section 4.2.1 about pulse EDMs.

Next we try to estimate the operating range of the EDM with a typical prism reflector. Here we specify the initial parameters of an EDM operating in this mode:

- The red laser luminous power F_{out} at the optical system output is 1 mW (10^{-3} W).
- The emission divergence Θ_l is 1 mrad.
- The receiving objective diameter D_{obj} is 50 mm ($5 \cdot 10^{-2}$ м).
- The single prism reflector diameter D_{ref} is 65 mm.
- The optical detector sensitivity F_{in} is $3 \cdot 10^{-11}$ W for red light
- The atmospheric transmittance τ^s is equal to 0.32 for red light at the distance of 5000 m.

Let us insert these values into Equation 4.14:

$$S = 0.8/\Theta_l \sqrt[4]{\frac{F_{out}\tau^{2s}D_{ref}^2 D_{obj}^2}{F_{in}}}$$

$$= 0.8 \cdot 10^3 \sqrt[4]{\frac{10^{-3}W \cdot 0.32^2 \cdot 42.25 \cdot 10^{-4} m^2 \cdot 25 \cdot 10^{-4} m^2}{3 \cdot 10^{-11} W}}$$

$$= 5537\ m.$$

We should correct the atmospheric transmittance value τ^s for this distance:

$$\tau^s = 0.8^{5.537} = 0.29$$

After inserting this value, a corrected S is equal to 5277 m. The maximal range for a single prism correlates to the manufacturer's specifications (about 5000 m for Sokkia SET530R).

4.2 TIME INTERVAL MEASURING METHODS APPLIED IN EDMs

4.2.1 Pulse Method (Time-of-Flight Method)

The main block of an EDM is the timer that measures discrete intervals with very high accuracy. In order to understand how this occurs we should examine the principles of its operation in detail. Nowadays two principles are applied in EDM distance measuring: direct pulse principle and shift-phase one. The structural scheme of an impulse EDM is shown in Figure 4.8.

Based upon commands from the microprocessor, the narrow pulse generator produces a very short electrical impulse. The duration is several nanoseconds. This pulse is transformed by the laser diode into a short light pulse whose power is several watts. Simultaneously with the light pulse generation, the timer starts up. Light is directed to the object by the means of the optical collimating system. A part of the reflected light power is collected by the receiving objective of the EDM. In the focus of the objective there is an APD that turns light pulses into electrical signals. At the photodiode output other than the valid signal, there is also a noise signal. The reason

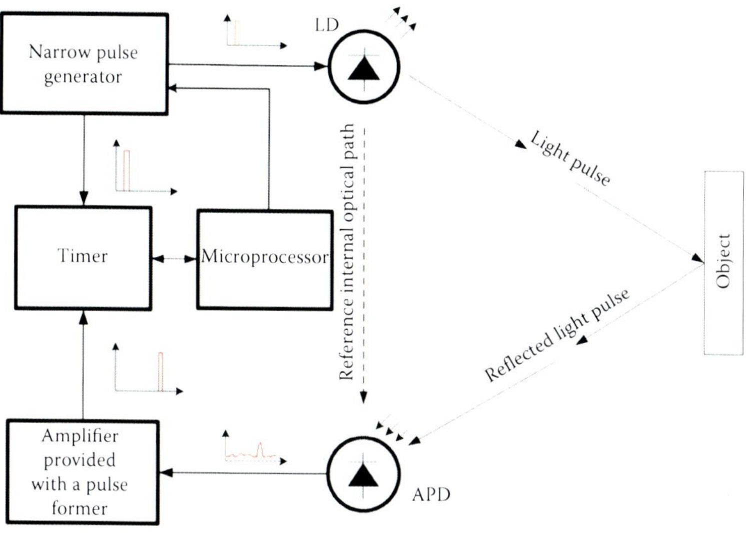

FIGURE 4.8 Pulse EDM principle.

for it is the photodiode's dark current and atmospheric light noise. The mix of the valid signal and noise enters first the amplifier, and then the former. Here the signal is amplified, filtered, and standardized. At the device output there is a pure pulse signal that stops the timer counting. Recently the timer has often been combined with a microprocessor. Nevertheless, the principle of its operation has not changed.

A typical timer scheme for a pulse-based EDM is illustrated in Figure 4.9. Measurable time intervals are generated by the means of an SR latch. This bistable element has two separate control inputs. When a pulse enters the S input, a logical 1 level is set at the output. If a pulse enters the R input, a logical zero level is set at the output. Thus, if the start pulse enters the S input and the EDM receiving channel pulse enters the R input, a positive pulse is generated at the SR flip-flop output. This pulse duration

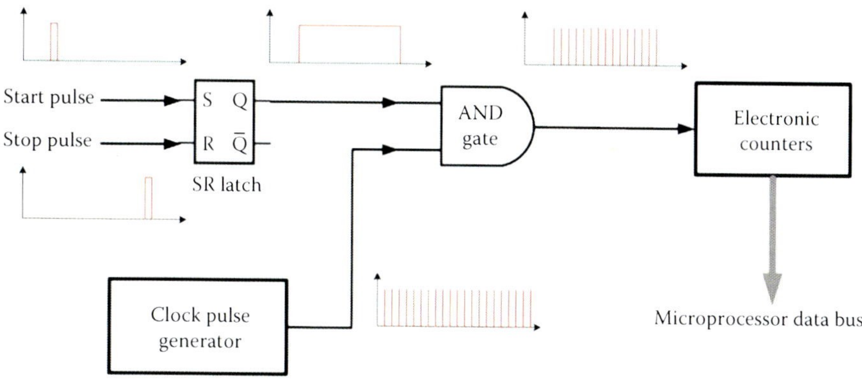

FIGURE 4.9 Digital timing.

corresponds to the total time of light pulses advancing a double distance and also electrical delays in the EDM electronic circuits. During the measurable time interval the counters accumulate clock pulses entering from the high-stable count pulse generator. A pulse packet is formed by the means of a logical element "AND gate." The element is able to produce a positive signal at the output only if there are positive signals at both inputs. Pulse packets are generated as its output enters electronic counters.

After finishing the count cycle, data regarding the pulse quantity enters the microprocessor via the data bus. Afterward, the microprocessor instructs to switch the optical channel into the internal reference path. Again a measuring cycle is carried out. The measurement result is deducted from the previous one. Thus, signal delays in the electronic circuits of the EDM are taken into account. Afterward this measured result is multiplied by the light's velocity value and enters into the display.

Distance measurement errors while using pulse EDMs may occur because of two main reasons. The first error is caused by discreteness of the timer operation (Figure 4.10). When the measurable interval is completed with clock pulses whose period is T_{clock}, there have always been the rest, t_q. The rest t_q maximal value could be close to the T_{clock}.

The timer speed depends both on the counters and the SR latch speeds. Maximal frequency of their performance is limited to about 1 GHz, which is equal to a time resolution of 1 ns. During this time, light passes the distance of 0.15 m twice. That means that during a single distance measurement applying the pulse method the

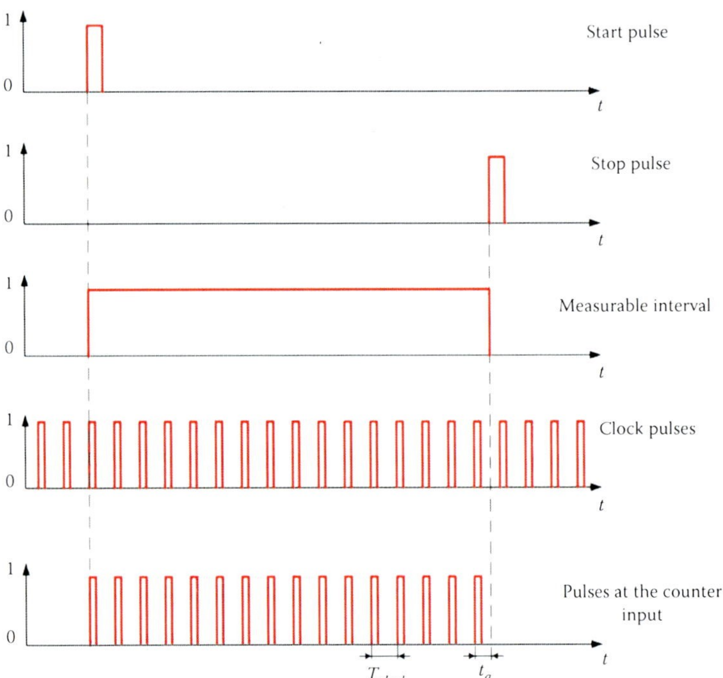

FIGURE 4.10 Pulse EDM discreteness.

counters discreteness is equal to 0.15 m. Such accuracy is not enough for surveying; however, impulse EDMs of this type do exist and land surveyors implement them for reconnaissance.

The discreteness effect can be minimized by averaging repeated measurements. For that, the start pulse is launched according to random distribution.

Another reason for error with pulse method is asymmetry of the pulse form. The actual form of the pulse signal at the photodetector output is in Figure 4.11 (the upper part). The pulse has a definite length of rise and fall. The signal at the photodetector output is a mix of pulses consisting of valid signals and noise. After amplification the mixed signal enters the former, which separates it from the noise. As the former, a threshold element is used. The threshold element operating level is higher than the noise level. If the input's amplitude changes, different time delayed pulses appear at the threshold's element output (see Figure 4.11, the lower part). The delay change may reach the pulse rise length. Contemporary EDM photodetectors are able to produce pulses whose rises are about 3–6 ns. That is why in such cases we use an elementary threshold element as the former distance and measurement error may

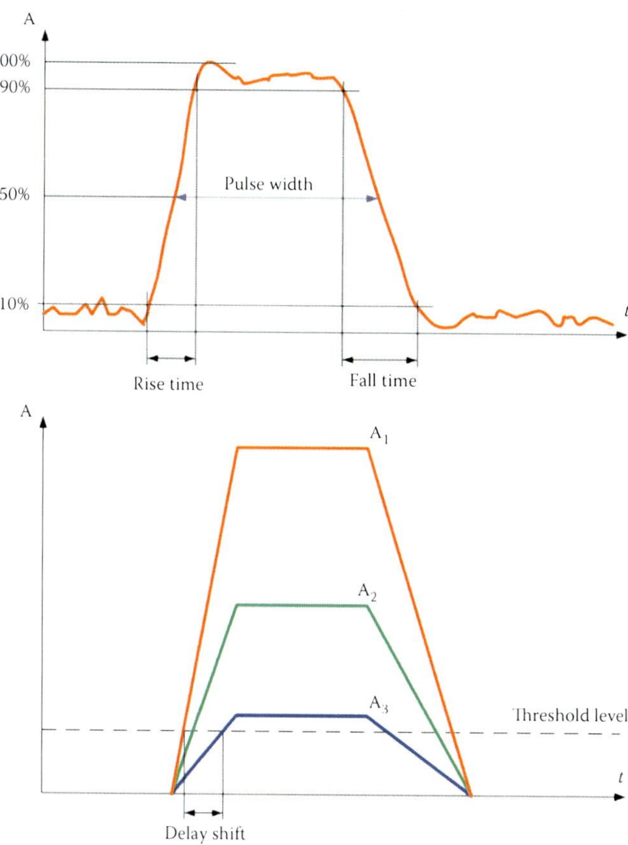

FIGURE 4.11 Pulse fixation error.

reach 0.5 m. The elementary method to decrease this error consists of finding the energy pulses center whose time position does not depend on the pulse amplitude. For the occurrences, the former uses the fall and rise pulses.

This method is effective only in case of equal duration for the both rise and fall of the pulse. Falls and rises of an actual pulse are not equal. That is why some of the amplifications' influence on the measurement results is present. In order to minimize this influence, the method is applied in conjunction with the pulse amplitude stabilization method.

Most pulse EDMs have pulse amplitude stabilizers where the pulse packet is integrated. These stabilizers have different outputs actuating the device, however their input devices are based on the same principle.

After preliminary amplifying, a part of the input signal enters the integrate device. At this device output there is a slowly changing signal that is equal to the signal averaged during a period of time. This time is called integration time. It varies from 5 to 500 ms, depending on the type of a stabilizer. Controllable variable resistors, or servomotors, are applied as actuating devices. The variable resistors are connected into the amplifier's signal circuit so that while increasing the signal at the amplifier, the input amplification factor decreases. This type of correlation is referred to as negative feedback. This technical solution performs well and allows decrease influence of atmospheric fluctuations. This solution's drawback is that the actuating device is set behind the photodetector. As a result the photodetector remains unprotected from strong optical signals. This occurs during short distances measurements or while pointing to a highly reflective object. That is why an optical attenuator is very often applied as an actuating device with the input amplitude stabilizer of an EDM (Figure 4.12). The stabilizer keeps a stable optical signal at the photodetector input.

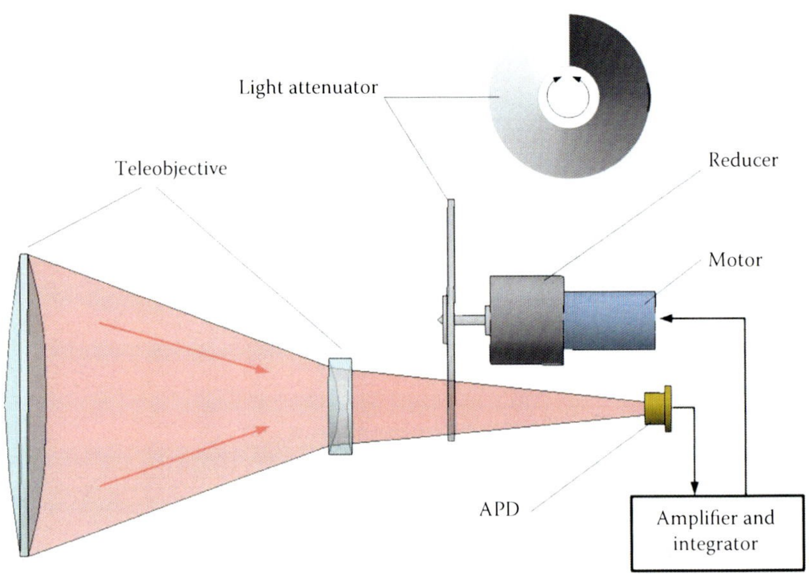

FIGURE 4.12 Amplitude stabilization method.

The signal level is preset so that the photodetector's valid output signal exceeds the noise level by a factor of several times.

An actuating device is a reverse electric motor including a gear reducer with an optical attenuator on its axis. In modern times, a stepper motor is applied instead of an electric motor. An attenuator is a variably transparent optical density disc. This disc is often called a gray wedge. If the optical input signal differs from the preset level, a regulating signal appears at the integrator's output. This signal makes the motor turn the disc so that optical input signal drifts to the preset level. Therefore, the power of light at the photodetector input is corrected.

The motorized actuating device has low performance and is not able to control atmospheric fluctuations of an optical signal. In order to solve the problem, an amplitude stabilizer, including the electronic actuating device described earlier, may also be applied. Either an amplifier having a controlled amplification factor or a bias voltage APD driver can be used as the actuating device.

In the case of an APD, using the avalanche multiplication factor is controlled. Stabilizers equipped with electronic actuating device have high enough processing speed to minimize atmospheric fluctuation influence. Nevertheless, electronic actuating devices suffer from this basic defect. That is why the stabilizers are used restrictedly, as their control elements operating in a great dynamic range may distort the time delay value. This is not acceptable for high precision EDMs.

Some techniques for minimizing time dependence on amplitude in pulse EDMs were adopted from experimental nuclear physics. One of these effective techniques is called amplitude and rise time compensated timing.

An input pulse signal is split into two channels (Figure 4.13). In one of the channels, input pulse delay occurs without changing the pulse amplitude. The delay value is set to be approximately equal to the pulse rise time. In the second channel, the input signal is reduced twice and inverted. Then signals from both channels are summarized. At the summated output we get a bipolar signal. The signal has a particular characteristic. There is a stable point on the signal curve whose time position does not depend on the signal's amplitude change (Figure 4.14). The signal enters the comparator, whose reference voltage is adjusted to this stable point. The pulse at the comparator output has a time-stabilized rise. If adjustment is correct, the technique allows minimized time dependence on the amplitude 100 times.

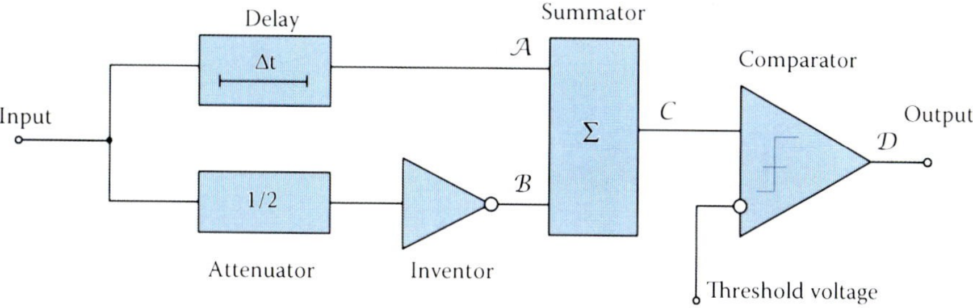

FIGURE 4.13 Amplitude and rise time compensated timing.

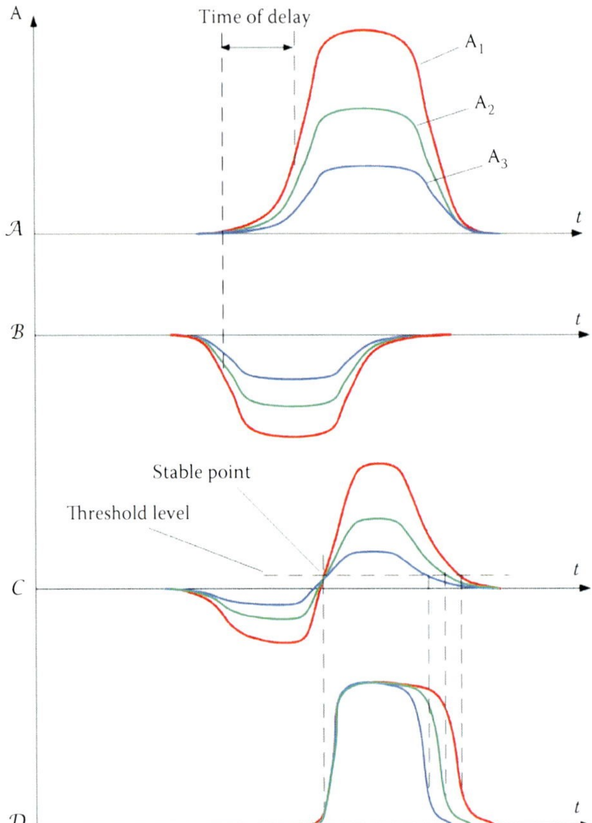

FIGURE 4.14 Signal waveforms at the points *A*, *B*, *C*, *D* shown in Figure 4.13.

Modern surveying pulse EDMs are able to measure lines with a high precision of 0.5–3.0 mm. Such results can be achieved only by using combined techniques of both optical and electronic signal amplitude stabilization. The stable points on the signal curve are used. As a rule, the measurement result averages 10^4 to 10^5 elementary measurements to reach a higher degree of accuracy.

In addition to these time proven methods, nowadays the WFD method is applied to improve the accuracy and the work range of a pulse-based EDM. This method improves the signal-to-noise ratio (SNR) and allows for working with signals whose amplitude is comparable to the amplitude of noise. The WFD method can be applied by using a high-speed analog-to-digital converter (ADC). Its maximum sampling rate can reach 1 GHz and its bandwidth is 300 MHz. One of the high-speed ADCs, the so-called parallel ADC, is shown in Figure 4.15.

A parallel ADC is also referred to as a flash ADC. This type of ADC has a large number of voltage comparators working together in parallel. The input signal enters the inverting inputs of the comparators at the same time. The reference voltage supplies the noninverting inputs of the comparators by means of resistor dividers. The reference voltage increases stepwise from the lower comparator to the upper one.

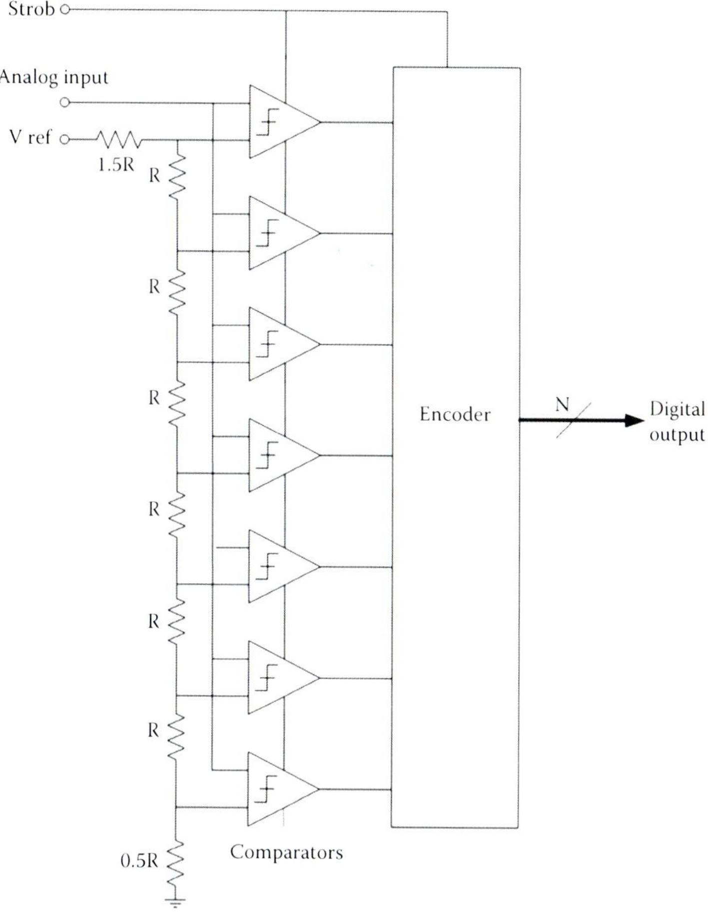

FIGURE 4.15 High-speed analog-to-digital converter (ADC).

The part of the comparators whose reference voltage is lower than an input level switches over, and their outputs become a logical 1. The outputs of other comparators are still equal to logical 0. At the outputs of the comparators, a digital code, which is called "a code of the mercury thermometer," appears. This code is inconvenient for microprocessor processing because a large number of communication lines are required. So it is usually transformed by an encoder to a convenient binary code that is put into the processor memory cells.

The transmitter of a pulse EDM emits very short pulses (up to 10 ns). The repetitive frequency of these impulses ranges from several kilohertz to several megahertz. The processor can change this frequency depending on the length of the measured distance and level of the reflected signal. Every single pulse is transformed by the means of the ADC into the data array (Figure 4.16). As we can see, the data arrays addresses are located along the time axis t and the value in each cell from this array correspond to the input signal level at the moment of the time quantization. As the

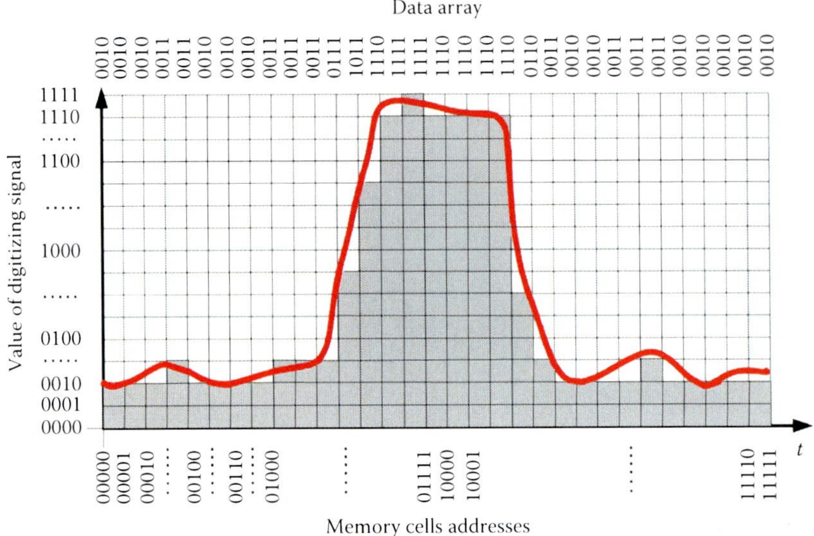

FIGURE 4.16 Digitalization of an analog signal.

single short pulses sequences entering the ADC have a determined accurate timing T, following digitalization they enter the same memory cells. This information is accumulated in the memory.

Noise signals have a casual amplitude and polarity. They are averaged while the data accumulates. At the same time the useful signal increases with the accumulation. A repeated digital accumulation improves SNR and allows using very weak signals and considerably increases the range of pulse EDM activity (Figure 4.17).

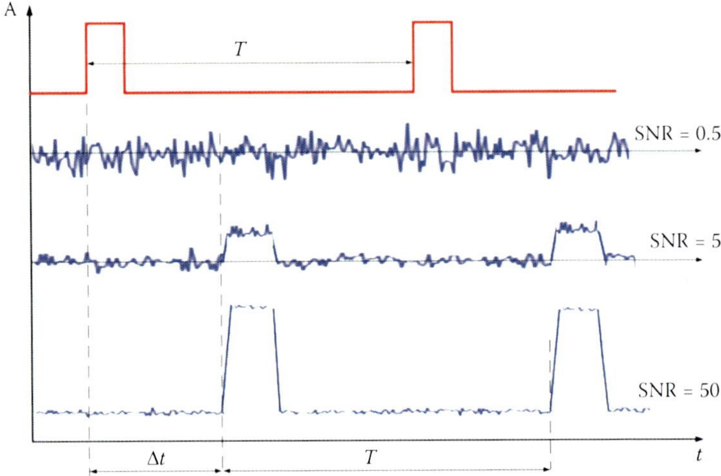

FIGURE 4.17 Result of waveform digitizing technology (WFD).

4.2.2 Phase-Shift Method

In phase EDMs (Figure 4.18) light passage time t of distance is determined by the phase shift $\Delta\varphi$ between periodic signals from transmitting and receiving channels:

$$t = T \cdot n + \Delta\varphi \cdot T/2\pi \tag{4.15}$$

where T is the high-frequency oscillator signal period, n is the number of full oscillation periods during the time interval t, and $\Delta\varphi$ is the phase shift (radian).

A high frequency oscillator generates a frequency-stabilized signal that enters the laser diode. This signal frequency is within 15 and 500 MHz. The laser sends amplitude-modulated light to the object. The optical signal covers there and back the measuring distance. At the photodetector output we receive a high frequency signal displaced in phase relative to the laser diode signal. Accurate direct measurement of these signal phase shifts is a problem that is difficult to solve. That is why the method of heterodyning is applied in surveying phase-shift EDMs. In the EDM there is one more frequency stabilized oscillator whose frequency is slightly different (several kilohertz) from the basic oscillator's frequency. This oscillator is called a heterodyne. If signals from both the heterodyne and the basic oscillator are sent to a nonlinear electronic device, according to the heterodyning principle, there is a mix of four frequencies at its output. This mix includes the signals from the basic oscillator, the heterodyne, beat frequency, and sum frequencies. At the mixer output there is a low frequency filter. As a result only the beat frequency signal (1–10 KHz) appears at the mixer output. The same mixer is available in the receiving channel. Here the heterodyne signal is mixed with the photodetector output signal. Finally there are two low frequency signals whose phase shifts can be accurately measured.

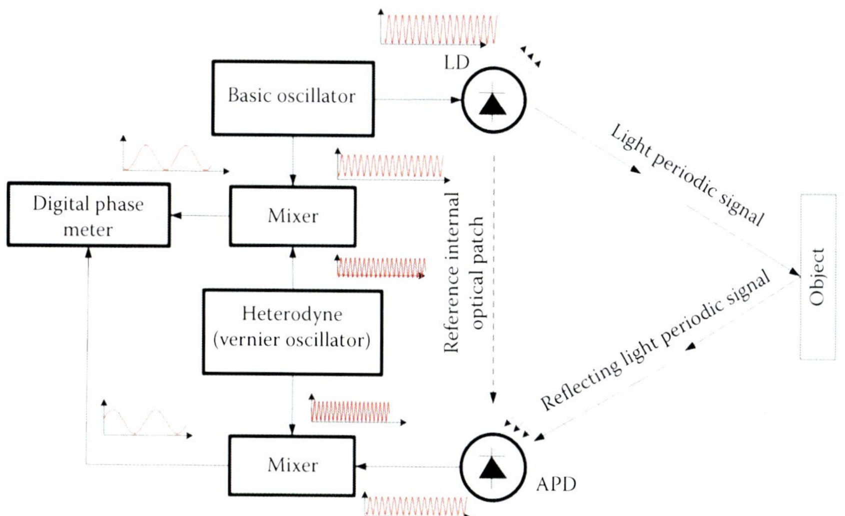

FIGURE 4.18 Phase-shift EDM principle.

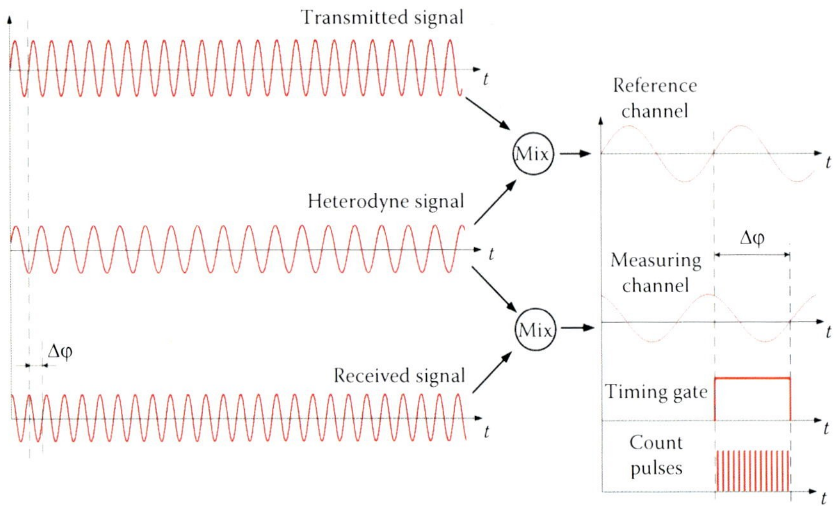

FIGURE 4.19 Heterodyne digital phase-shift measurer.

During heterodyning the phase shift is the same as the initial high frequency signal phase shift (Figure 4.19). A digital phasemeter is the same digital measurer of time intervals (see Figure 4.10), but its timing gate is formed not by a pair of start–stop pulses but rather the signals from both transmitting and receiving channels mixers.

Also, it is necessary to find out the number n of full phase cycles (or the signal periods) according to Equation 4.15. In contemporary phase-shift EDMs the problem is resolved through the s method of phase-shift measuring at several frequencies. As a rule there are three of them. The first is a basic one, and the two others are auxiliary frequencies. They are several times lower than the basic frequency. The lowest frequency signal period is chosen so that time of light passage for the longest measuring distance should be less than this period. Taking into account Equation 4.13 we can set up a three-equation system. Here the unknown values are t, n_1, n_2. The given values are signal periods T_1, T_2, T_3; and $\Delta\varphi_1$, $\Delta\varphi_2$, $\Delta\varphi_3$ are measured phase shifts of three frequencies. The number of equations and unknown values is equal. This system of equations is solved as

$$\begin{cases} T_1 \cdot n_1 + \Delta\varphi_1 \cdot T_1/2\pi = t \\ T_2 \cdot n_2 + \Delta\varphi_2 \cdot T_2/2\pi = t \\ \Delta\varphi_3 \cdot T_3/2\pi = t \end{cases} \tag{4.16}$$

Data from the digital phasemeter enters the EDM microprocessor, which handles the processing of the equation system.

Use of two frequencies is enough for portable handheld EDMs whose maximum range is within hundreds of meters.

In order to exclude time delay drift there is an internal reference optical path in phase-shift EDMs. Phase shift measurement is fulfilled both in distance measurement mode and the internal optical path measurement mode. The data enters the EDM microprocessor, which calculates the distance and adds atmospheric correction data to it.

4.3 SEMICONDUCTOR PHOTODETECTORS AND LIGHT SOURCES OF EDMs

4.3.1 General Principle

Nowadays only semiconductor photodetectors and emitters are used in surveying instruments. Next we will investigate the principles of their workings.

Semiconductors are matter- or chemical-based compounds taking up a position between insulators and electric current conductors. Matter conductivity depends on the atom's outer shell electron conditions. Metal atoms are arranged in a crystalline structure bound by valent electrons. A considerable part of the metal's outer shell electrons are not involved in such a bond and form an intercrystalline electronic gas. It contributes to their high conductivity.

An insulator's outer shell of electrons is firmly bound with atoms or involved in interatomic bonds. In such a case the conductivity is very little since free charge carriers are absent.

Admittedly, conductivity depends not only on the atom's properties but also on its bond types. An impressive example of this is carbon, which can exist as diamond or graphite. In standard conditions, diamond is an insulator, whereas at high temperature it acts as a semiconductor. Graphite conductivity is more like a conductor. We can see that conductivity greatly depends on temperature.

Silicon's use as a semiconductor is often applied in modern electronics. The reason for this is because of silicon's wide distribution in nature. Silica sand entirely consists of silicon oxide. A silicon atom is shown in Figure 4.20.

In a silicon crystal lattice, silicon atoms are bound in pairs with adjacent atoms by two common valent electrons. At very low temperatures, free charge carriers are

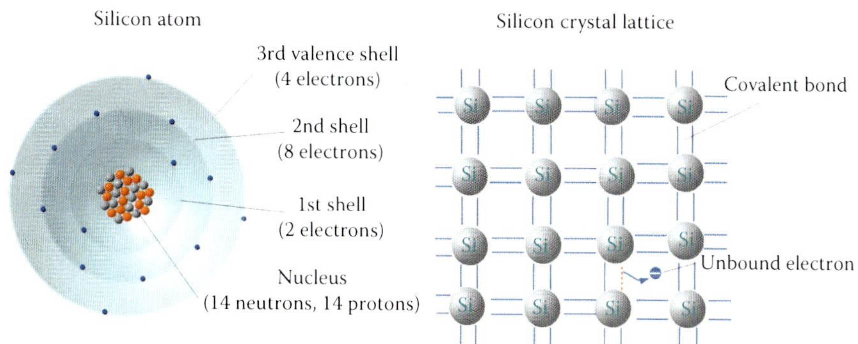

FIGURE 4.20 Silicon atom structure.

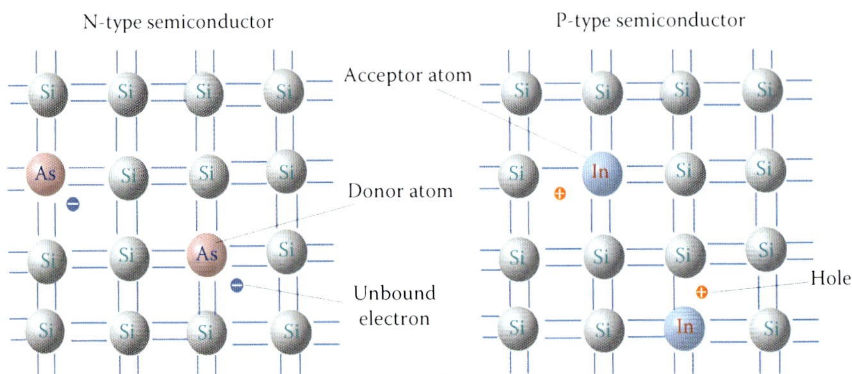

FIGURE 4.21 Types of semiconductors.

absent in the crystalline material. In such conditions, crystal silicon becomes an insulator. Whereas at a normal temperature, the electron's thermal mobility reaches such an extent that several electronic bonds are torn. The electron has become free and is able to move in interatomic space. If the electron leaves its position, a vacant electric-positive point appears. In electronics this vacant place is called a hole. A free electron could bind with the hole over time. This process is called recombination.

The number of free carriers in a semiconductor can be increased if impurities have been introduced into it (Figure 4.21). If an impurity containing a fifth valent electron (arsenic, for example) is introduced into a silicon crystal lattice, four covalent bonds are enabled. Meanwhile, the fifth electron becomes free. Conductivity caused by electrons prevails in semiconductors of this type. Semiconductors of this type are called N-type semiconductors.

If the impurity atoms (indium, for example) have only three valent electrons, then electron holes may appear in the semiconductor lattice. Semiconductors of this type are called P-type semiconductors. If voltage is applied to such a semiconductor, these vacant places are filled with adjacent electrons that vacate the next point. Current caused by moving positive holes occurs in the material.

Contact of two semiconductors of different types may create very remarkable characteristics (Figure 4.22).

Such a contact is called the *p-n* junction, and a device where this contact is physically carried out is called a semiconductor diode. In the *p-n* junction area electrons from a N-type semiconductor position themselves into vacant lattice holes of P-type semiconductor. This means that electron and hole recombination occurs. Such a process takes place only within a narrow area of a *p-n* junction. This area becomes depleted of free current carriers and an electrical field preventing further advancement of the current carriers occurs. This field is created by a positive and negative lattice of ions in the semiconductor, emerging after recombination. This field is called built-in potential. If forward bias is applied to the *p-n* junction more than the built-in potential, a current runs through the *p-n* junction. In the case of a reversed application to the *p-n* junction, the insulating area is extended, and the current is

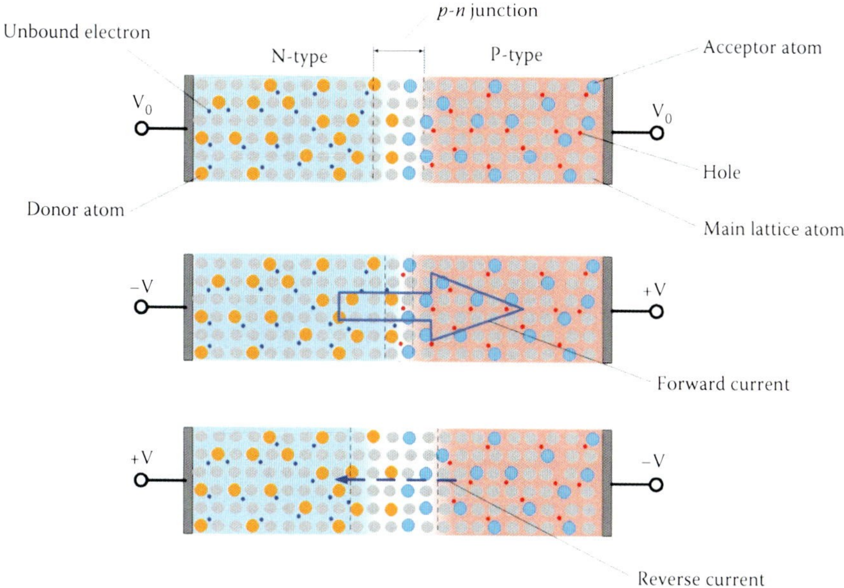

FIGURE 4.22 Semiconductor p-n junction.

practically absent. Now it is necessary to make an important remark. Insignificant current through the p-n junction still runs. The reason for this is the fact that a N-type semiconductor has, besides electron conductivity, scanty P-type conductivity, and so in a P-type semiconductor there is a scanty quantity of free electrons. These current carriers are referred to as a minority. They are exactly the reason for the existence of current through p-n junctions when a reversed bias is applied.

If a reversed bias p-n junction is lit a reversed current increases (Figure 4.23). This is caused by the fact that photons are absorbed by electrons that move to their next power level and are able to leave their places in the crystal lattice. In other words, due to the influence of light, an additional quantity of free carriers may appear in a semiconductor.

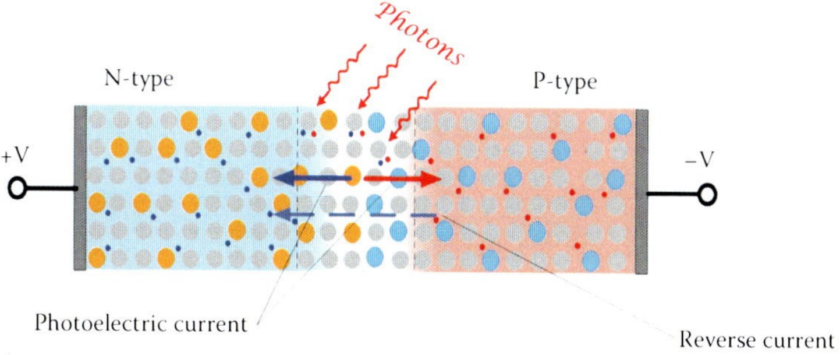

FIGURE 4.23 Light influence upon the semiconductor p-n junction.

4.3.2 Photodiodes

A specialized semiconductor diode for light detection having a window to illuminate the *p-n* junction is called a photodiode. A reversed current through a photodiode in the absence of light is a parameter restricting the photodiode sensitivity to small light powers. This parameter is called a photodiode dark current. A reversed current through a photodiode induced by light is called a photocurrent.

Ordinary photodiodes are not applied in EDMs because of their low sensitivity and insufficient performance. In order to minimize a photodiode, dark current special PIN photodiodes are worked out (Figure 4.24, upper part). PIN photodiodes have a thin layer of poor extrinsic semiconductors (*i* – layer) in the *p-n* junction area. *p-n* junction components themselves have a high concentration of impurity.

Making an avalanche photodiode has become the next step in the development of semiconductor photodiodes. A semiconductor photodiode structure is generally similar to PIN photodiode structures. However, in an avalanche photodiode structure between a high-alloy area having electron conductivity and *i* – layer there is a lay with a regular concentration of acceptors (Figure 4.24, the lower part). In an avalanche photodiode, the avalanche effect occurs. This effect is created when electrons impact atoms at high speeds, and these electrons are able to knock additional electrons out of them. This effect is possible only at a high acceleration voltage. So an avalanche photodiode reverse bias is usually within 100 and 400 V. Every free electron appearing in *i* – layer because of the photo effect is able to produce the appearance of several hundreds of derivative free electrons as a result of the avalanche effect.

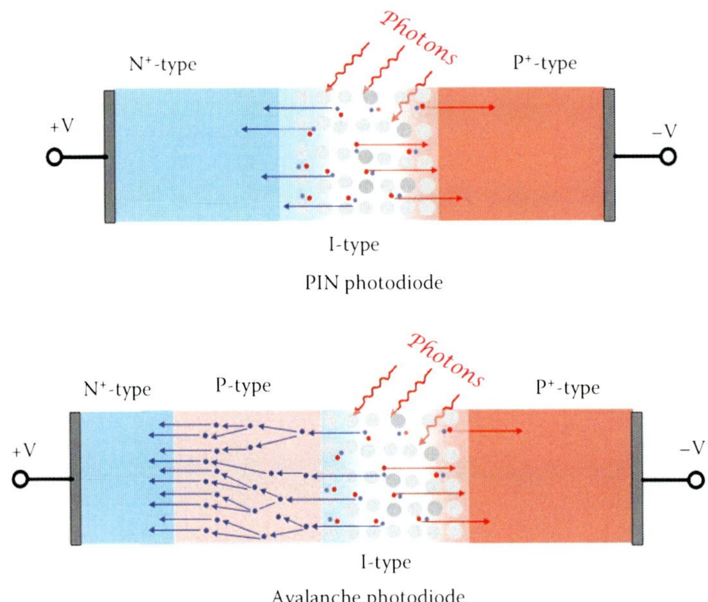

FIGURE 4.24 Types of photodiodes.

In modern surveying EDMs, only avalanche photodiodes are used as photodetectors. The best of them are able to detect light at the level of several photons, and their performance is less than one nanosecond. Their disadvantages are high cost, and due to the avalanche multiplication factors, dependence on temperature. Recently, specialized microchips have been created for power supply of avalanche photodiodes. The chips are equipped with temperature sensors and a high-voltage regulator at the output. When using a power supply chip, an avalanche photodiode reverse bias is changed so that during temperature change, the avalanche multiplication factor of the instrument is still invariable.

4.3.3 Light-Emitting Diodes and Lasers

Now let us consider applying a *p-n* junction as a source of light. This case is more difficult. Any *p-n* junction can be used as a photodiode. The question is what is the quality of the photodiode. Meanwhile, not every *p-n* junction can be forced to emit light power.

During direct application of voltage to *p-n* junctions, a current runs through it, and the recombination of electrons and holes occurs in a narrow area of the *p-n* junction. This recombination is attended with electron transition to a lesser energy state. In a silicon crystal lattice, such an energy jump occurs with insignificant energy release. In silicon all energy is spent to the thermal vibrations of the lattice atoms. Such a type of recombination is called nonemitting. The chemical compound GaAs is also a semiconductor. In it an electron loses considerable energy during recombination. Such an energy jump may produce a photon of light. This type of recombination is called emitting. In the early stage of GaAs light-emitting diodes, luminescence was very weak, due to the fact that a considerable part of the recombination was nonemitting. One of the reasons for this is that the recombination took place in a wide area. The area was successfully narrowed by applying a *p-n* junction consisting of similar but still different chemical compounds. The *p-n* junction of this type is called a heterojunction. A modern light-emitting diode structure is shown in Figure 4.25.

Infrared light-emitting diodes started to become more widespread in the 1960s. As a result of this, EDMs started shifting from field laboratories into compact

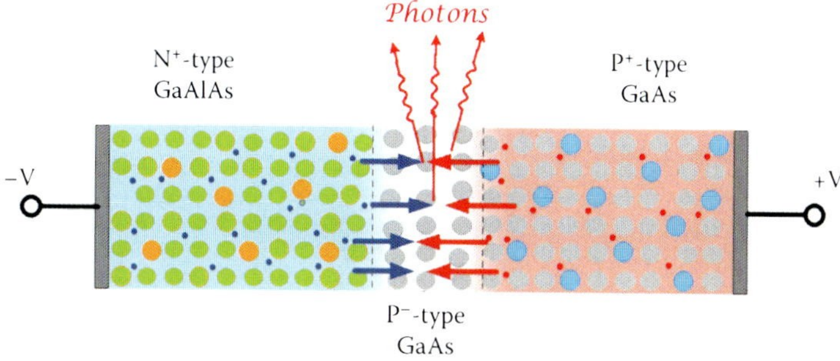

FIGURE 4.25 Heterojunction light-emitting diode.

instruments. More often we would see them as an optical theodolite attachment and also as parts of the first total stations.

The appearance of semiconductor lasers became the next step in furthering the development of light-emitting diodes.

Electron transmission into a lesser energy state in a concrete substance causes a light photon of a defined wavelength. If the transmission is synchronized with other electron transmissions we can get a narrow spectrum range laser emission. In order to create it, we put the emitting medium into an optical resonator. Two polished facets of some *p-n* junction material are used as such a resonator. The nonemitting recombination should be minimal in semiconductor lasers. The first semiconductor lasers could only operate in a pulse mode and with an intense level of cooling. As soon as the slightest superheating occurred, they would begin emitting spontaneous emission and they turned into ordinary light-emitting diodes or were even destroyed. Later, double heterojunctions started to be used. This resulted in a significant increase of emitting medium quality and allowed for the design of lasers with the ability to operate in a nonstop mode. Such a laser structure is demonstrated in Figure 4.26.

Semiconductor laser power depends on the heterojunction temperature. When the temperature increases, emission power decreases. In order to control emission power, a photodiode sensor is set up next to the heterojunction. The current that supplies the laser is controlled by a special driver whose input is connected with the photodiode sensor. The dispersion diagram of the semiconductor laser has an ellipse shape.

From the beginning of the 1980s to the beginning of the 1990s infrared lasers were the primary light sources used in surveying EDMs. At the beginning of the 1990s red semiconductor lasers appeared. As a result, a new range of portable handheld laser EDMs came to existence. At the end of the 1990s visible red lasers began appearing in total station EDMs. Currently, most total stations are equipped with these EDMs. Nevertheless, infrared lasers are still used. Most silicon avalanche

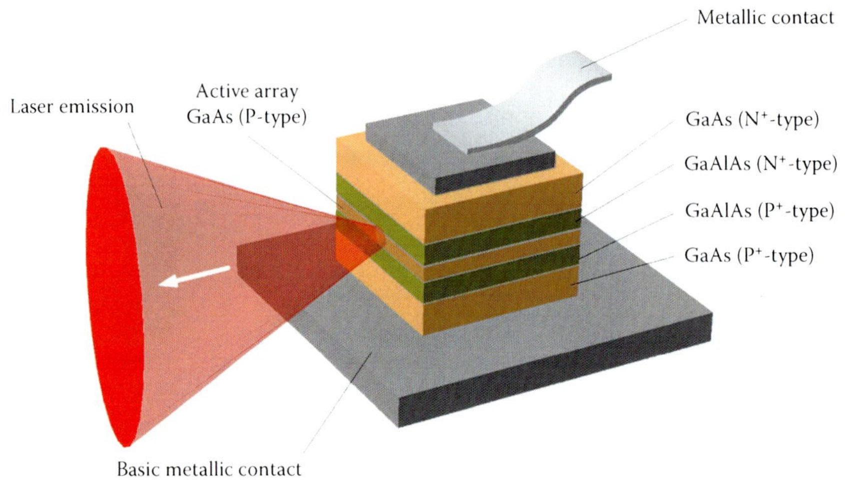

FIGURE 4.26 Semiconductor laser structure.

photodiodes are more sensitive in the infrared range (about 0.8 mkm). Also, as noted earlier, the atmosphere is more transparent in this optical range. That is why in the design of maximal range surveying EDMs infrared lasers are mainly used.

4.4 EDM OPTICAL SCHEMES

As stated earlier, the first surveying EDMs were developed as separate instruments and only lately have they been integrated into total stations. As a separate device, an EDM is used only as a portable handheld rangefinder. Instruments of this type are used not only by land surveyors but also builders. A typical portable handheld EDM is displayed in Figure 4.27. A basic monocomponent receiving objective and an outlet laser window are arranged on the front facet of the instrument case. The EDM's upper board has a large LCD display and operation keys. Using them we can switch on/off the instrument, start up the measurement mode, and perform some arithmetical calculation of the measurement results. Various makes may have additional devices such as circular or cylindrical fluidal levels, sighting devices, and a compass.

An elementary optical scheme consisting of separated optical receiving and transmitting channels is applied in these EDMs. This technical solution allows the receiving objective to collect the maximum volume of reflected light power. An elementary monocomponent objective also provides a maximal transmitting efficiency and minimal cost. The monocomponent objective application became possible due to the narrow spectrum light used in this type of EDM. This allows us to disregard chromatic aberrations in the optical channel. The spherical aberrations issue with a monolens objective is solved by means of reasonably long focus lenses (up to 150 mm) whose diameters are small (up to 30 mm). Nevertheless, a noncoaxial optical scheme has several unwanted effects. First, it is parallax. The principle of the parallax influence is the following (also see Figure 4.28).

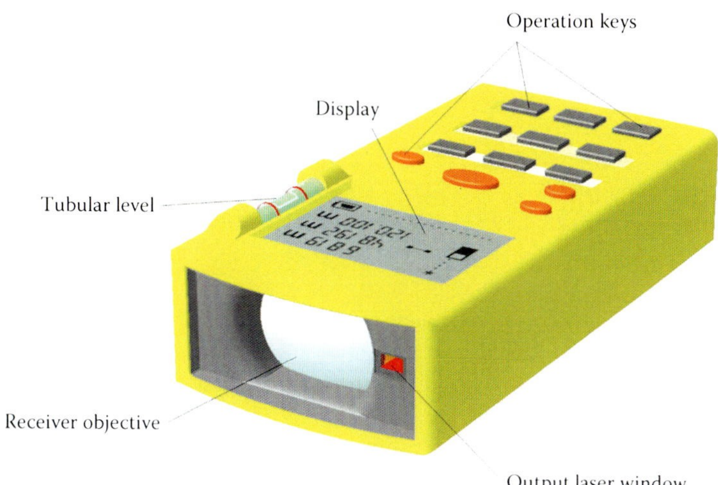

FIGURE 4.27 Portable laser rangefinder.

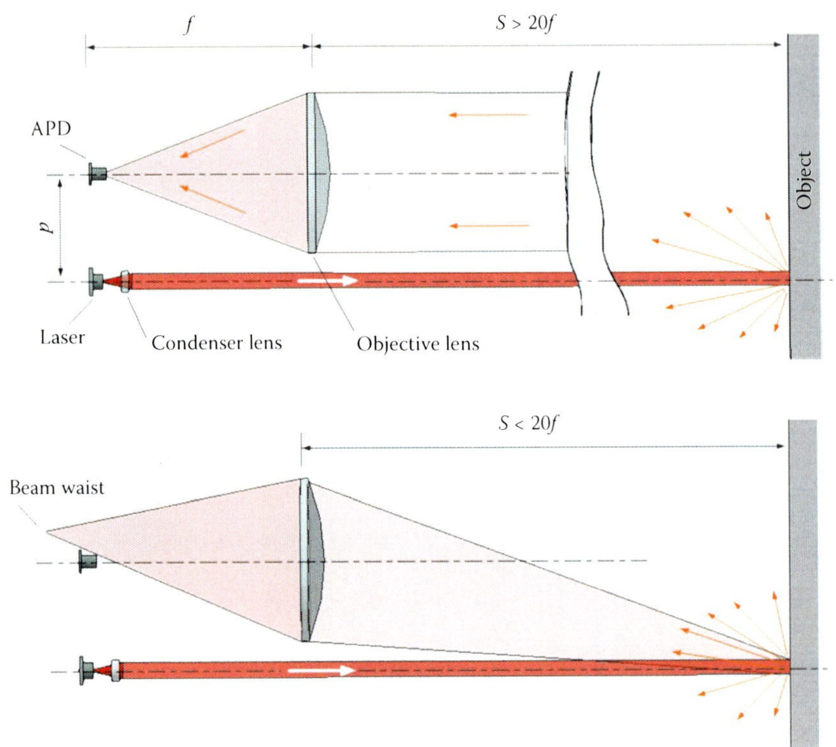

FIGURE 4.28 Ray path in a portable laser distance meter.

The laser spot image projected to the object is found aside from the receiving objective axis. As a result, the receiving objective creates the spot image in the point, which also is not present on the objective axis. Nevertheless during long-distance measurements, deflection of the laser spot image from the objective axis is very small, so light enters the photodetector. During short-distance measurements the receiving objective makes the spot image aside and behind the photodetector. Starting from a certain distance the reflected light stops entering the APD.

In the very first portable handheld EDM ("Disto" Leica) the problem of parallax was addressed efficiently and simply. Optical fiber was introduced into the optical scheme. One end of the fiber was set in front of the APD, and the other end was placed on the focal surface of the receiving objective, itself having been fixed on a mobile bracket. The motor axis had an eccentric cam and controlled movement of the bracket. The motor received instruction from the EDM microprocessor. The microprocessor also controlled the fibers end movement and found an optimal position of the fiber relative to the objective axis.

Now the influence of parallax is minimized with the inclusion of additional optical components within the EDM's optical scheme (Figure 4.29). Various components such as optical wedges, cylindrical lenses, or mirrors may be used. During short distance measurement, a small portion of the light rays is turned toward the

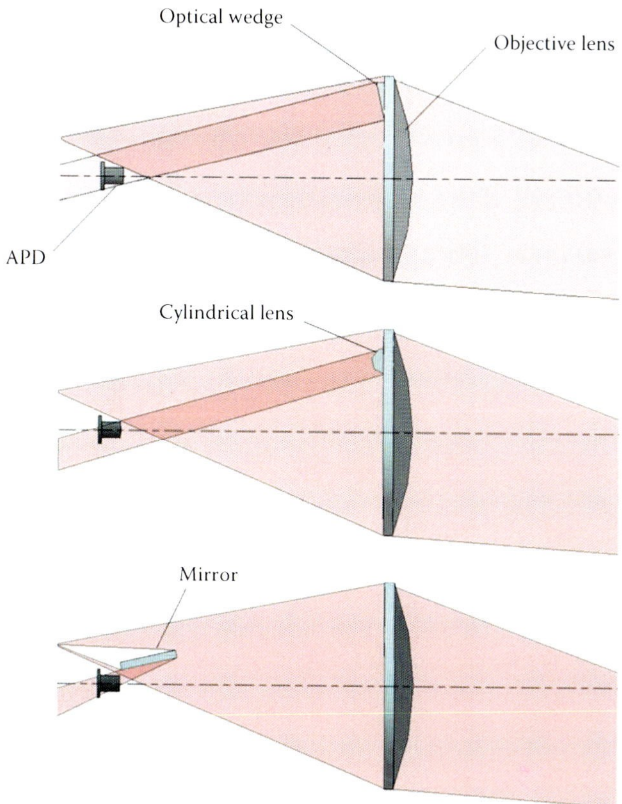

FIGURE 4.29 Parallax influence correction.

photodetector by the means of these components. Cylindrical lenses and wedges are placed on a small area of the receiving objective to create minimal shading within the receiving objective. A mirror does not at all influence amount of light received in an optical signal. These additional optical components are made of minimal size to minimize background exposure, which negatively influences upon a photodetector sensitivity. Let us take a look at the typical optical scheme of a hand portable EDM (see Figure 4.30).

The transmitting channel consists of a red laser diode, a condenser lens, and a partially reflecting mirror. The mirror is almost transparent and reflects a very small amount of light power. Usually it is less than 1%. Such a mirror uses a transparent glass plate without an antireflection layer. The mirror is set up at a certain angle toward the laser beam and directs a small part of the light power to the photodetector. Therefore, a reference internal light patch is formed.

The input optical channel is more complicated. The receiving objective's area is made as large as possible. Usually this is an edged lens whose diameter is from 4 to 8 cm. This objective occupies the largest part on the front side of the EDM. The input end of the optical fiber is placed in the receiving objective focus. The optical fiber is

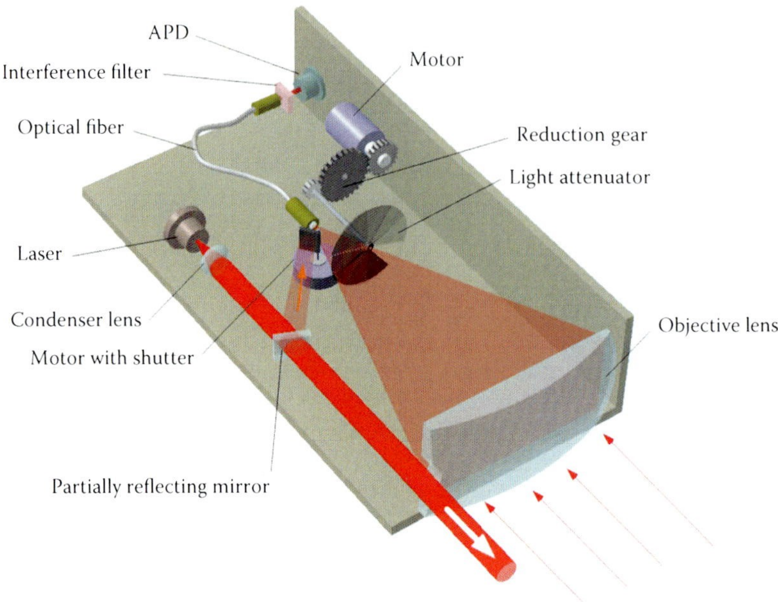

FIGURE 4.30 EDM optical scheme where a gray wedge is used.

a piece of glass or plastic fiber protected with a solid shell and is plugged into holders (Figure 4.31). The holders and fiber ends are usually carefully polished. There is a red interferential light filter at the fiber output. It admits only a narrow optical range and thus reduces the photodetector background light. The interferential light filter is effective only for parallel rays, therefore a condenser lens is set in front of it. The second condenser lens is set behind the light filter and collects parallel rays in the light-sensitive area of the APD. Some APDs have a collective lens at the input. In this case, the second condenser lens may not be required.

Often the optical fiber is intentionally bent at several points. This contributes to better mixing of light in the fiber and more even light distribution at the fiber output.

There is a variable transparency optical filter between the receiving objective and the input end of the fiber. It is often called a gray wedge. It is made as a disk

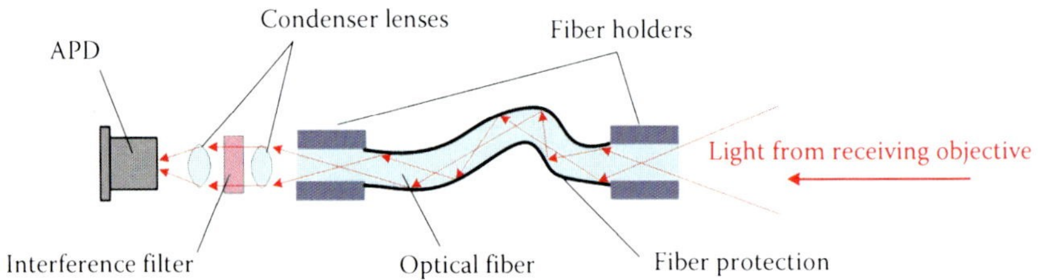

FIGURE 4.31 EDM light detector with an optical fiber.

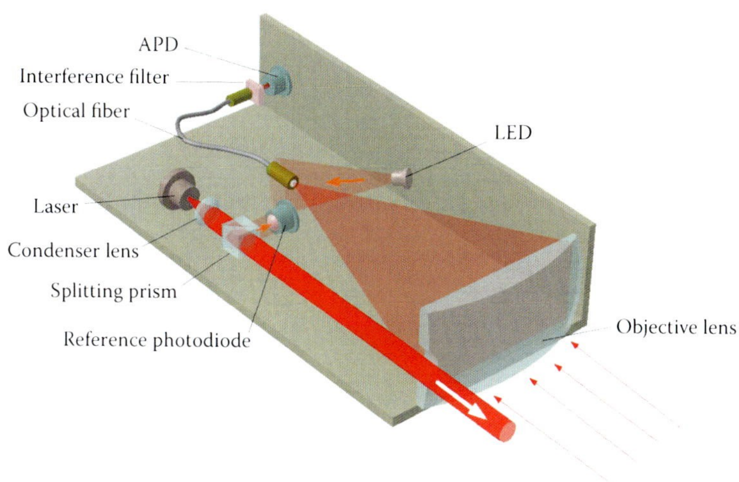

FIGURE 4.32 EDM optical scheme without a gray wedge.

adhered to the axle of the gear reducer. The device's purpose and structure is similar to an executive device for optical signal stabilization, shown previously in Figure 4.12.

In order to commutate measuring and reference channels, there is an electric motor with a shutter attached to its axle. The shutter is rotated by the motor within a limited rotary angle. In the measuring-to-object mode, the shutter shuts the light that the partially reflecting mirror sends to the optical fiber input. In the reference internal distance measuring mode, the shutter shuts the light coming from the receiving objective and lets the light come from the partially reflecting mirror. The optical scheme shown in Figure 4.30 is common for many types of up-to-date surveying EDMs. Many of the components of the scheme are applied in EDMs of total stations. Nowadays this optical scheme can be considered as a classical one.

Another optical scheme for portable handheld EDMs is shown in Figure 4.32. It is applied only in handheld EDMs, most notably the portable Leica models.

The main difference in this scheme from the classical one is that an optical switcher of reference and measuring channels is not present. This is possible because of use of one more receiving channel. For that, a part of the laser emission is deflected with the help of a beam-splitting prism toward the second photodiode. It makes no sense to use an expensive avalanche photodiode, as the photodetector of the reference channel for optical signal levels from the beam-splitting prism is enough for a high-speed PIN photodiode. The EDM processor simultaneously processes the data from both the measuring and reference channels.

Another significant difference is the absence of a gray wedge in front of the photodetector. A light-emitting diode (LED) is set in the sight of the avalanche photodiode. It is used for faint illumination of the avalanche photodiode. By changing LED emission we can control the avalanche photodiode's sensitivity and stabilize the signal level in the measuring channel.

TABLE 4.2
Modern Laser Distance Measurers

Model	Maximal Measure Distance (m)	Accuracy (±mm)	Vertical Angle Sensor (Accuracy±°/ Range°)	Display	Memory (Result/ Photo)	Interface	Manufacturer
S910	300	1	0.1/360	Color touch	50/80	Bluetooth WLAN	Leica Geosystems
D810 touch	200	1	0.1/360	Color touch	30/80	Bluetooth	Leica Geosystems
D510	200	1	0.2/360	Color	30/—	Bluetooth	Leica Geosystems
LD-520	200	1	—/360	Color	30/—	Bluetooth	Stabila
GLM250VL	250	1	—	Monochrome	30/—	—	Bosch
QM95	200	1	—	Monochrome	—	—	Spectra Precision
LD-500	200	1	—/90	Color	20/—	—	Stabila
D410	150	1	—	Color	30/—	—	Leica Geosystems
X310	120	1	0.2/360	Monochrome	20/—	Bluetooth	Leica Geosystems
PD-512N	120	1.5	—	Color	99/—	—	South
D3aBT	100	1	0.2/360	Monochrome	20/—	Bluetooth	Leica Geosystems
GLM100C	100	1.5	0.1/360	Monochrome	50/—	USB 2.0	Bosch
Fluke 424	100	1.5	0.2/360	Monochrome	—	—	Fluke

Apparently, this special optical scheme does not have mobile optical and mechanical junctions. However, in the designing of EDMs as parts of total stations, Leica developers still apply a classical optical scheme that apparently provides higher accuracy, ranging, and temperature rates.

The most well-known laser distance measurers are listed in Table 4.2.

BIBLIOGRAPHY

Books and Reports

Kester, W. 1996. *High Speed Design Techniques*. Norwood, MA: Analog Devices.

Maar, H. and H.-M. Zogg. 2014. WFD—Wave Form Digitizer Technology. White paper. Leica Geosystems AG.

Rüeger, J. M. 1996. *Electronic Distance Measurement*. Berlin: Springer-Verlag.

Patents

Chaborski, H. 1980. Laser distance measuring apparatus. US Patent 4,181,431 filed October 5, 1978, and issued January 1, 1980.

Ehberts, H., H. Bernhard, K. Giger, and J. Hinderling. 1998. Device for distance measurement. US Patent 5,815,251 filed May 4, 1994, and issued September 29, 1998.

Gogolla, T., A. Winter, and H. Seifert. 2006. Laser distance measuring device with phase delay measurement. US Patent 7,023,531, B2 filed August 8, 2003, and issued April 4, 2006.

Hildebrand, K. 1959. Method and device for electro-optical distance measurement. US Patent 2,909,958 filed November 21, 1956, and issued October 27, 1959.

Hildebrand, K. 1960. Electro-optical distance meter. US Patent 2,956,472 filed October 16, 1957, and issued October 18, 1960.

Howard, B. R. and K. D. Froome. 1970. Distance measuring apparatus which compensates for ambient atmospheric refractive index. US Patent 3,521,956 filed November 9, 1964, and issued July 28, 1970.

Liu, H.-T., H.-Q. Chen, and L.-B. Yu. 2008. Laser distance measuring system with a shutter mechanism. US Patent 7,471,377 B2 filed July 30, 2007, and issued December 30, 2008.

Liu, Y. 2008. Range finder. US Patent 2008/0007711 A1 filed March 23, 2006, and issued January 10, 2008.

Mira, S. and R. Schwarte. 1989. Optoelectric distance measuring apparatus with delay and zero cross detector. US Patent 4,849,644 filed August 5, 1988, and issued July 18, 1989.

Oishi, M. and Y. Tokuda. 2009. Processing apparatus for pulsed signal and processing method for pulsed signal and program therefore. US Patent 7,518,709 B2 filed January 28, 2005, and issued April 14, 2009.

Rudolf, W. K. 1975. Device for electro-optical distance measurement. US Patent 3,898,007 filed November 27, 1972, and issued August 5, 1975.

Schwarte, R. 1988. Optoelectric distance measuring apparatus with e time discriminator for the accurate detection of electric pulse sequence. US Patent 4,734,587 filed January 22, 1986, and issued March 29, 1988.

Siefert, H., M. Penzold, U. Krueger, and G. Schusser. 2001. Laser distance-measuring instrument for large measuring ranges. US Patent 6,281,968 B1 filed October 25, 1999, and issued August 28, 2001.

Siercks, K. 2009. Optical distance measuring method and corresponding optical distance measuring device. US Patent 2009/0262330 A1 filed July 13, 2007, and issued October 22, 2009.

Tiedeke, J. 1992. Pulse delay measuring circuit. US Patent 5,102,220 filed November 6, 1990, and issued April 7, 1992.

5 Total Stations

A total station is a measuring instrument applied to measure distance, and vertical and horizontal angles. This is a tool that combines the functions of a theodolite and an electronic distance measurer (EDM), and has a microprocessor with software.

The first electronic total station, Reg Elta 14, was presented by the West German Zeiss in 1968 (Figure 5.1). Prior some leading companies that produced surveying instruments combined an optical theodolite or electronic theodolite with an independent EDM. Nevertheless there was no calculator in them or even if there was a calculator routine manual input of angular measurements results was required. This combined device was the prototype of an actual total station. However, the EDM and the theodolite forming the prototype were independent units. The Reg Elta 14 and subsequent total stations had a considerable difference: the telescope of the theodolite and the EDM optical part became one unit.

Reg Elta 14 created a great furor at the Olympic Games in Munich in 1972. In full view of the whole world on Olympic stadium an unprecedented apparatus on a tripod, similar to a TV camera, measured sport results instead of people with habitual measuring tapes. The result was immediately relayed to the nearest big tableau.

Reg Elta 14 was an impressive size of 560 × 310 × 340 mm and weighed about 20 kg. Its distance range was 500 m in a single prism mode, and up to 2000 m using a 19-prism reflector. The horizontal angular accuracy range was 10″ and the vertical angular accuracy range was 15″. A portable perforator was applied as a recorder. A paper tape was used as data storage. Then, information from the data storage was loaded into a stationary computer. First, a total station allowed one to fulfill surveying without a handwritten register. Reg Elta 14's effectiveness was quickly estimated at true worth and soon almost all leading surveying instrument manufacturers worked out similar total stations. Thus a new era in surveying began.

Current total stations are of the same concept (Figure 5.2). The basis of a total station is an electronic theodolite. This kind of a theodolite was reviewed in Chapter 3. We should point out that operation of the theodolite part of a total station is identical, in principle, to an electronic theodolite. Many manufacturers apply their electronic theodolites as the main elements in total station development. The main difference between an electronic theodolite and a total station is in their telescope design. Total station telescope have a built-in EDM. We will discuss it in detail in this chapter. There are some other differences between these instruments. A total station computer is more powerful than an electronic theodolite calculator. The tasks that a total station fulfills require a more advanced display. The most advanced total station has a graphic display with a touch screen. This kind of display is not applied in electronic theodolites. Also a total station has a more powerful battery, as its EDM requires more energy.

Later a new type of a total station appeared called a motorized total station. It was equipped with motive-powered drivers. Their appearance allowed one to carry

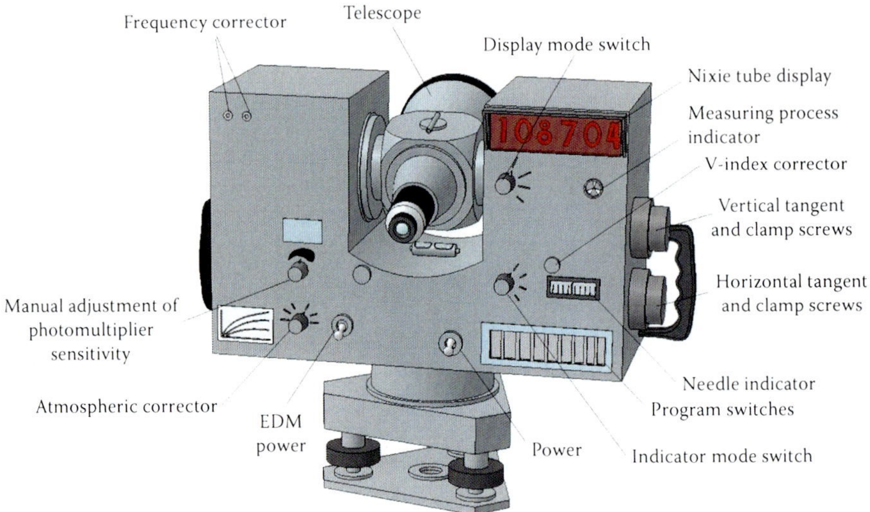

FIGURE 5.1 Total station Reg Elta 14.

FIGURE 5.2 Routine total station.

out automatized measurement such as autolock. Autolock allows fulfilling of semi-automatic measurements. Such total station locks on to an "active" reflector, then it precisely follows as the reflector moves from point to point.

The next advancement was the appearance of the robotic total station. Current robotic total stations have more rapid drivers such as servo drivers or piezoelectric drivers. The robotic total station allows a single operator to perform all necessary measurements using automatic aiming to a reflector and radio communication between the device and its controller. The controller of the device is set on a landmark with an "active" reflector and provides full control under the device when the operator is in a measured point.

Robotic total stations (Figure 5.3) do not usually have the prototype theodolite, since it is not possible to place a mechanical drive into an electronic theodolite's main body. Nevertheless, the robotic total station theodolite part operates on the same principle as an electronic theodolite.

Physical principles of EDM operation are reviewed in Chapter 4. The electronic part of a total station EDM and a "detached" EDM are similar, but the optical part is significantly different. The total station EDM optical part must be built into the telescope.

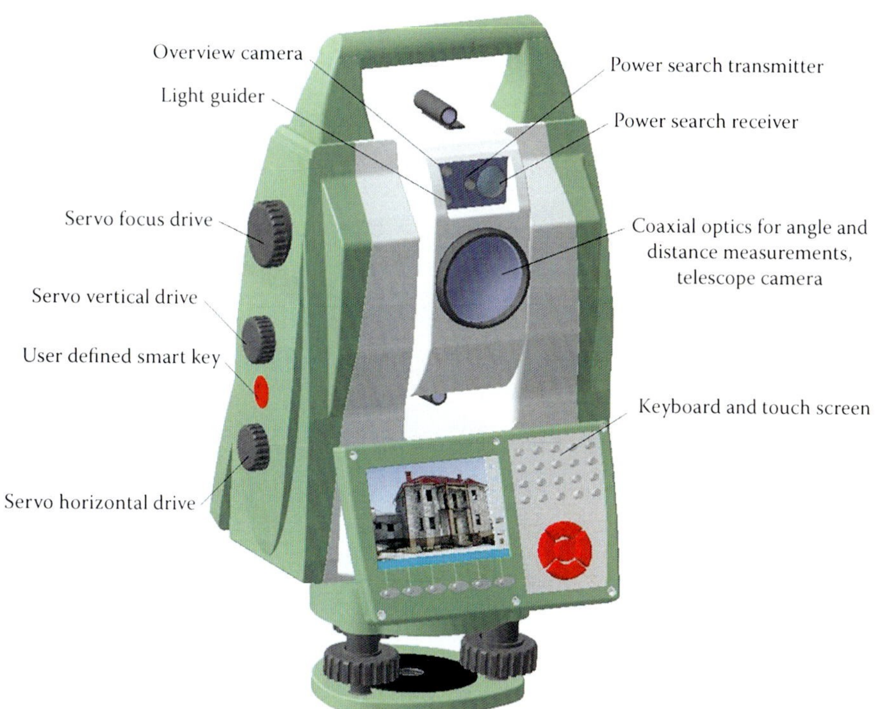

FIGURE 5.3 Robotic total station.

5.1 TOTAL STATION CONFIGURATION

5.1.1 Basic Axes of a Total Station

The basic configuration of a total station is shown in Figure 5.4.

The angle measuring device of a total station should meet certain requirements resulting from its basic axial configuration:

- The total station vertical axis should be set into the vertical position before measuring
- The total station horizontal axis should be perpendicular to its vertical axis.
- The collimation axis of the telescope should be perpendicular to the total station horizontal axis.

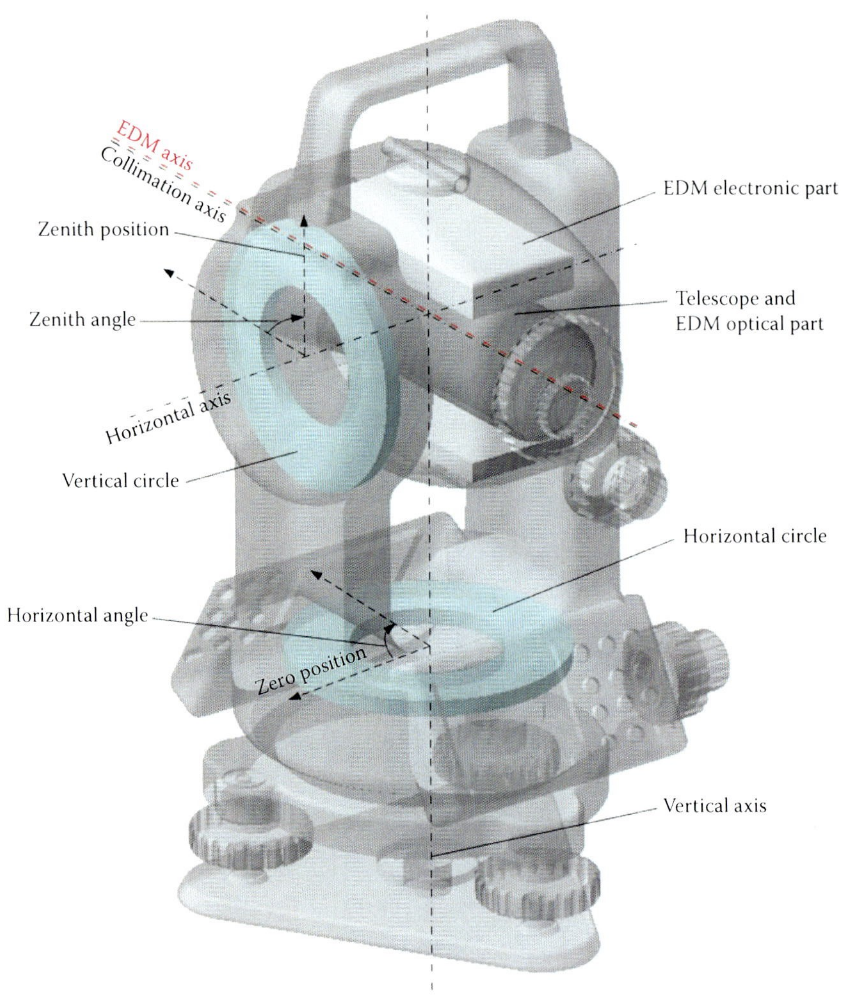

FIGURE 5.4 Basic axes of a total station.

The EDM device should meet one more requirement: *The EDM axis should be superposed with the collimation axis of the total station*. Thus, a total station consists of an electronic theodolite (reviewed in Chapter 3) and an EDM (whose principle of operation is described in Chapter 4). That is why the main issue considered in the following sections is the total station telescope.

5.1.2 Total Station Telescope

Total station EDMs are divided into three main groups:

- Infrared EDM requiring a reflector prism
- Reflectorless EDM with a red laser
- Reflectorless infrared EDM

5.1.2.1 Infrared Electronic Distance Measurer (EDM) Requiring a Reflector Prism

The first type of total station EDM includes an EDM that operates only with a prism or a special reflecting plate. The optical scheme of its telescope is in Figure 5.5.

The telescope consists of the sighting and the EDM part. The sighting part of the telescope consists of the same elements of an ordinary theodolite. There are the compound objective, a focusing lens, an inverting Porro prism, a reticle, and a compound eyepiece. Instead the Porro prism an Abbe prism may be used. The objective is common for sighting and EDM parts. The EDM optical part consists of a splitting dichromatic prism, a prism with mirror faces and with a correcting prism, protecting screens, and transmitting and receiving units. The transmitting unit has an infrared

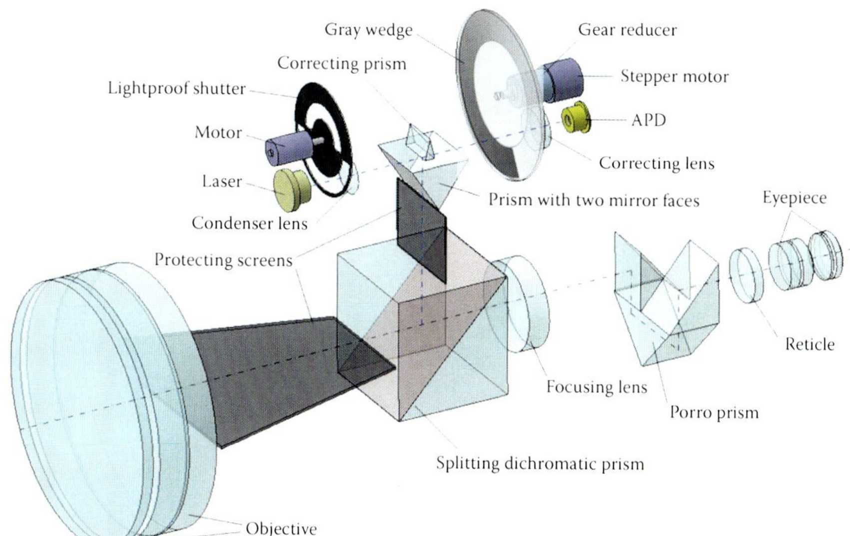

FIGURE 5.5 Optical scheme of the telescope of a total station with an EDM operating only in reflector mode.

laser supplemented with a condenser lens, and a motor with a shutter attached to its axis. The receiving unit includes an avalanche photodiode (APD), a correcting lens, and a stepping motor with a gear reducer whose axis has a gray wedge.

The splitting dichromatic prism divides the sighting and the EDM channels. The prism is made so that its inner face with a special covering reflects infrared light and freely transmits visual range light. Infrared laser is the source of light of this EDM. First, the laser light is collimated by the condenser lens and then enters the mirror faces prism. This prism has two light-reflecting faces and is placed so that it divides the field of vision of the objective into two equal parts: receiving and transmitting. In order to divide the EDM channels, additional screens are set. A reflecting prism is set at the other end of the measured line. The light that is sent back from this reflecting lens is collected by the receiving part of the objective and enters the splitting dichromatic prism. Then this light, reflected from the inner face of the prism, is turned by the mirror prism and enters the photodetector, which consists of an avalanche photodiode and a correcting lens. A gray wedge is placed between the mirror prism and the photodetector to keep the light at a constant level at the photodetector input.

Additionally, the EDM optical scheme has an inner calibrating channel. This channel operation is shown in Figure 5.6. On the left size of the figure, the laser light goes to a distant reflector. In this case the light goes through the external (measurement) channel. On the right size of the figure, the laser light is directed to the APD. The distance is measured in each of these two modes. The inner (calibration) channel length allows us to calculate the influence of the drift time delays of the EDM.

The inner and external channels are switched by a sector lightproof shutter placed onto the motor axis. This shutter has radial slots located on different sectors and at different distances from the center (see Figure 5.5). The motor movement is limited by two discreet positions. In case the motor turns the shutter so that the bigger radius slot is located opposite to the laser, the EDM measures distance to the reflecting prism (see the left side of Figure 5.6). In case the shutter is turned to the opposite position, the laser light does not enter the mirror prism and runs to the correcting prism and then enters directly the photodetector (see the right side of Figure 5.6). Despite the fact that total stations with such a telescope are able to work only using prism reflectors or reflecting plates, some manufacturers still produce these total stations (see Table 5.1).

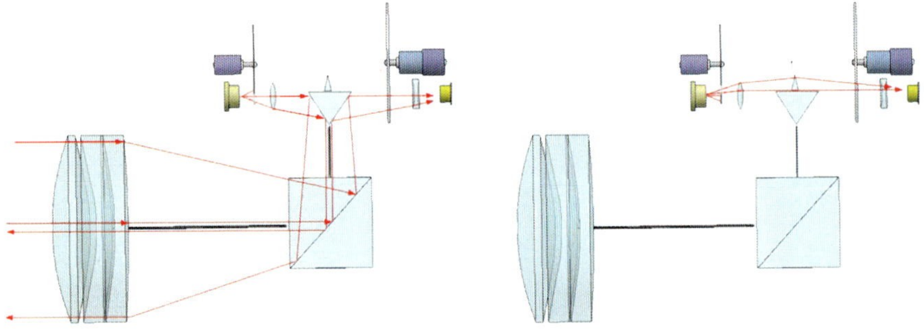

FIGURE 5.6 Ray tracing in the EDM shown in Figure 5.5.

TABLE 5.1
Up-to-Date Total Stations (without Reflectorless Mode)

Model	Angle Measure Accuracy (″)	Distance Measure Accuracy (mm)	Max. Measure Distance Mini Prism/ Single Prism/ Triple Prism (km)	Memory (1×10^3 points)	Interface	Manufacturer
DTM-322+	2(Hz)/5(V)	±(3 + 2 ppm × D)	1.0/2.0/2.6	50	RS-232C	Nikon
GTS-252	2	±(2 + 2 ppm × D)	0.9/2.0/2.7	24	RS-232C	Topcon
GTS-255	5	±(2 + 2 ppm × D)	0.9/2.0/2.7	24	RS-232C	Topcon
RTS-102	2	±(2 + 2 ppm × D)	—/3./—	120	USB, RS-232C	FOIF
RTS-102	5	±(2 + 2 ppm × D)	—/3.0/—	120	USB, RS-232C	FOIF

5.1.2.2 Reflectorless EDM with a Red Laser

The EDMs of the second group are able to operate both in reflector and in reflectorless mode. There are two types of telescopes in a reflectorless EDM depending on the location of the gray wedge. The telescope of the first type is in Figure 5.7. There is gray wedge in front of the photodetector (APD).

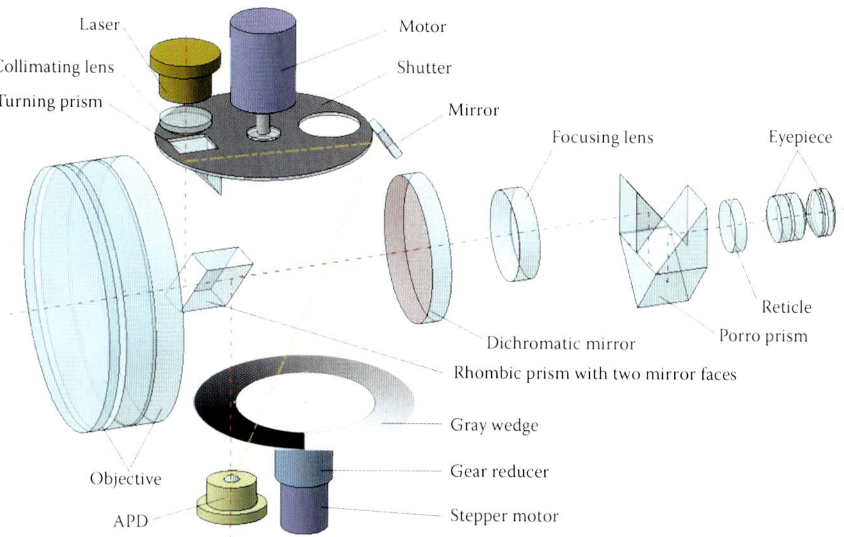

FIGURE 5.7 Optical scheme of the telescope of a total station with a reflectorless EDM (the first type).

The sighting part of the total station optical scheme is made of the same elements described earlier. The sighting part includes a composed objective, a focusing lens, an inverting Porro prism, a reticle, and a composed eyepiece. Instead of the Porro prism an Abbe prism may be used. The objective is also common for sighting and EDM parts. The EDM optical part consists of a prism with two mirror faces, a dichromatic mirror, and transmitting and receiving units. The transmitting unit has a red laser supplemented with a collimating lens, a motor with a shutter attached to its axis, and a mirror. The shutter has a turning prism to form a reference line. The receiving unit has an APD and a stepper motor with a gear reducer whose axis has a gray wedge.

A plane-parallel glass plate with dichromatic mirror coating on its one side is used as a divider between the sighting and the EDM channels. This coating reflects the light within a very narrow spectrum of wavelength of the laser emission. Red lasers are more frequently applied in these EDMs. Use of the nearest infrared diapason laser is infrequent. When using a red laser, the image quality strongly depends on the dichromatic mirror workmanship. The mirror must effectively reflect a very narrow red diapason light. If the diapason is a little wider, a considerable part of the visible spectrum does not pass through the plane-parallel plate, and the image seems to be colored in blue and red light appears as dark red.

The EDM receiving and the transmitting channels are divided by the means of a rhombic prism whose two faces have a mirror coating.

By means of the collimating lens, the laser emission narrows into a 5 mm visible laser beam. This beam enters the rhombic mirror prism and then goes through the objective to the object. In some reflectorless EDMs this mirror prism is available in front of the objective, but this does not change the principle of the optical scheme operation.

The light reflected from the object is collected by the objective and then the dichromatic mirror reflects it onto the back face of the rhombic mirror prism. Then the light goes through the gray wedge and enters the photodetector.

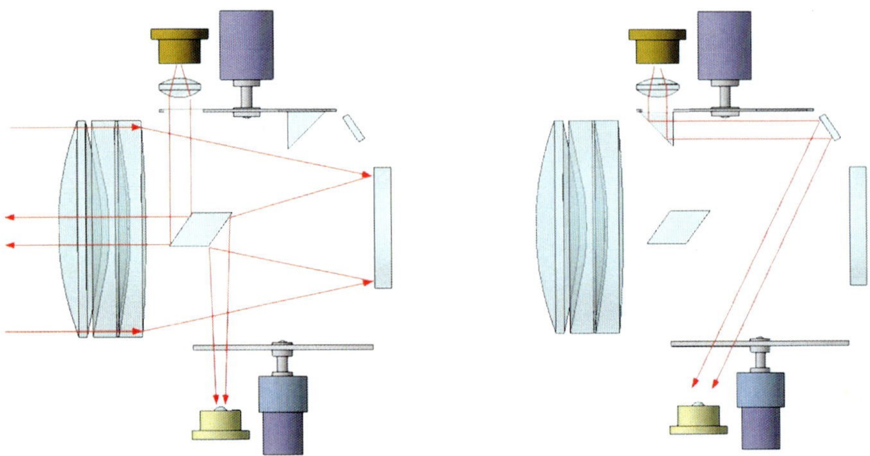

FIGURE 5.8 Ray tracing in the EDM shown in Figure 5.7.

Switching in the mode of the inner (calibration) channel is carried out by the rotating of the shutter on the motor axis. The motor run is limited within a half-turn. When we measure the distance to the object, the laser light goes through the hole on the shutter (see the left side of Figure 5.8). The inner (calibration) channel is measured by turning the shutter 180°. In this case, the laser light goes through the opposite hole in the shutter. There is the turning prism on the shutter behind this hole (see the right side of Figure 5.8). The light goes through the prism and enters the mirror that directs the light to the photodetector (APD).

The second type of telescope in a reflectorless EDM is shown in Figure 5.9. There is gray wedge right behind the laser. The sighting part is the same as in the first type of telescope. The transmitting part of the EDM is different. The laser light from the condenser lens goes through the gray wedge. The gray wedge on the disk is rotated by a stepper motor at an angle when the light power entering the APD keeps a constant level. The motor with its shutter is located between the gray wedge and the condenser lens. The motor operates as a switcher of the inner (calibration) and the external (measurement) channels. An additional dichromatic filter is located in front of the photodetector to protect it from background light.

The ray tracing in these channels are shown in Figure 5.10. The laser light goes through the gray wedge and enters a distant reflector when the shutter of the switcher is horizontal (see the left side of Figure 5.10). When the shutter is turned at about 45° the laser light directly enters the photodetector barring the objective (see the right side of Figure 5.10). Thus, switching in calibration measurement mode occurs. The

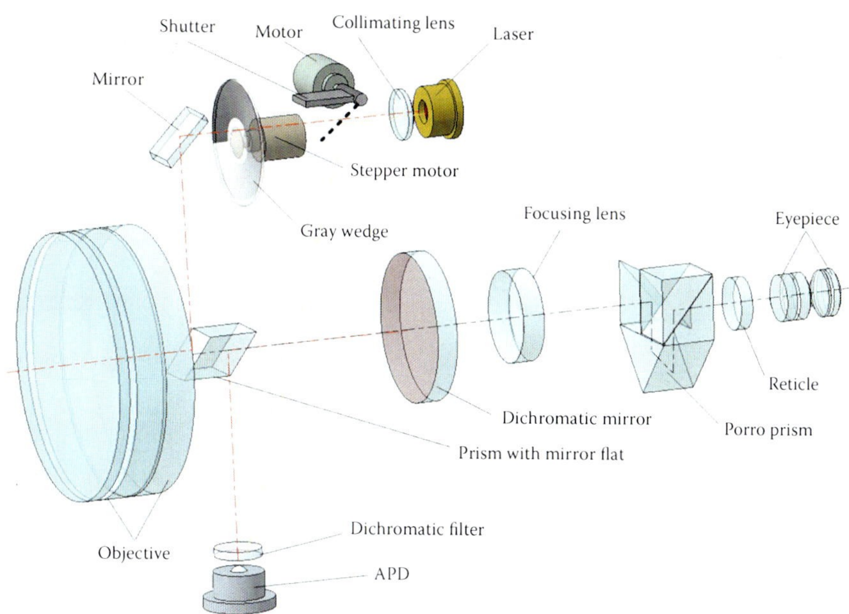

FIGURE 5.9 Optical scheme of the telescope of a total station provided with reflectorless EDM (the second type).

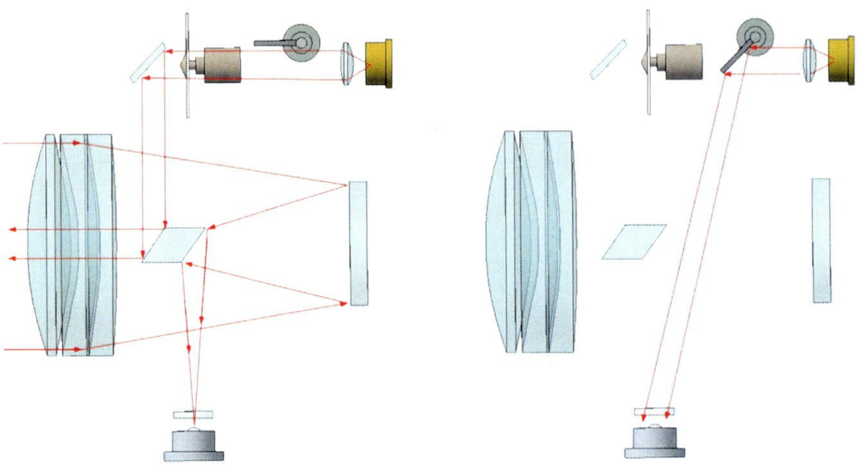

FIGURE 5.10 Ray tracing in the EDM shown in Figure 5.9.

shutter reflecting surface is rough, not mirror, as the calibration channel requires a little part of a laser light.

In the preceding we described the basic principle of a reflectorless EDM operation. The actual optical design may be different. It may have a fiber optic in the receiving channel, additional light filters placed in the front of the photodetector, and gray wedges to adjust the light in the calibration channel. These EDMs are also able to measure distance in reflector mode. In order to use the EDM in this mode, a light-reducing element is set into the transmitting channel. The element may be a lightproof shutter with a slot, a negative lens, or a dark glass.

Sokkia and Trimble apply the telescope of the first type. Leica, Pentax, South, and Foif apply the telescope of the second type.

Table 5.2 shows the reflectorless total stations made by leading manufacturers.

5.1.2.3 Reflectorless Infrared EDM

An unusual type of total station EDM is produced by Nikon. For about 10 years Nikon has applied it in its reflectorless total stations (Figure 5.11). Nevertheless, EDMs of this type are still widely used in total stations. The sighting optical part does not differ from a theodolite telescope and consists of a compound objective, a focusing lens, an inverting Porro prism, a reticle, and a compound ocular. However, the EDM part significantly differs from the two aforementioned configurations. The dividing dichromatic prism is placed not in the front of the focusing lens of the telescope but behind it. As a result the EDM and sighting channel focusing occurs simultaneously. The laser is placed into a focal plane of the teleobjective, so with careful focusing the laser spot on the object shrinks to its minimal size. Since a full teleobjective aperture is applied, this spot has a small size. The laser spot on the object is built up by convergent beams. This is impossible in the two preceding total station schemes, as their laser beams are almost parallel. A small size of the laser spot let us fix the object more exactly while surveying. An infrared

TABLE 5.2
Up-to-Date Reflectorless Total Stations

Model	Angle Measurement Accuracy (″)	Distance Measurement Accuracy (mm) Prism/ Reflectorless	Maximal Measurement Distance: Reflectorless/ Single Prism (km)	Memory (1×10^3 points)	Interface	Manufacturer
Builder 500	5, 9	±(2 + 2 ppm × D)/ ±(3 + 2 ppm × D)	0.25/3.5	50	RS-232 Bluetooth USB	Leica Geosystems
TS02 plus R500	3, 5, 7	±(1.5 + 2 ppm × D)/ ±(2 + 2 ppm × D)	0.5/10	24	RS-232	Leica Geosystems
TS06 plus R1000	2, 3, 5, 7	±(1.5 + 2 ppm × D)/ ±(2 + 2 ppm × D)	1.0/10	100	RS-232 Bluetooth USB	Leica Geosystems
TS09 plus R1000	1, 2, 3, 5	±(1.5 + 2 ppm × D)/ ±(2 + 2 ppm × D)	1.0/10	100	RS-232 Bluetooth USB	Leica Geosystems
TS11 plus R1000	1, 2, 3, 5	±(1.5 + 2 ppm × D)/ ±(2 + 2 ppm × D)	1.0/10	Windows 1Gb	RS-232 Bluetooth USB	Leica Geosystems
CX-100	2, 3, 5, 6	±(2 + 2 ppm × D)/ ±(3 + 2 ppm × D)	0.5/5	10	RS-232 USB	Sokkia
FX-100	1, 2, 5	±(2 + 2 ppm × D)/ ±(3 + 2 ppm × D)	0.5/5	MAGNET 0.5Gb	RS-232 Bluetooth USB	Sokkia
M3 DR	1, 2, 3, 5	±(2 + 2 ppm × D)/ ±(3 + 2 ppm × D)	0.5/5	Windows 1Gb	RS-232 Bluetooth USB	Trimble
R-1500	2, 3, 5	±(2 + 2 ppm × D)/ ±(3 + 2 ppm × D)	0.5/3	60	RS-232 USB	Pentax
W-1500	1, 2, 3, 5	±(2 + 2 ppm × D)/ ±(3 + 2 ppm × D)	0.5/3	Windows 1Gb	RS-232 USB	Pentax

laser is applied in this optical scheme. The laser spot is situated exactly in the reticle crosshairs. Its size is from one and a half to two bisectors of the reticle. The reflector EDM operates in two modes: to-object distance measurement and inner calibration (Figure 5.12).

On the right side of Figure 5.12 we can see ray tracing in to-object distance measurement mode. The most complicated and important optical element is a splitting dichromatic prism. It not only divides the sighting and EDM channels of the telescope but also integrates the transmitting and receiving channels of the EDM. The beams both of the transmitting and receiving channels are refracted thrice in this prism.

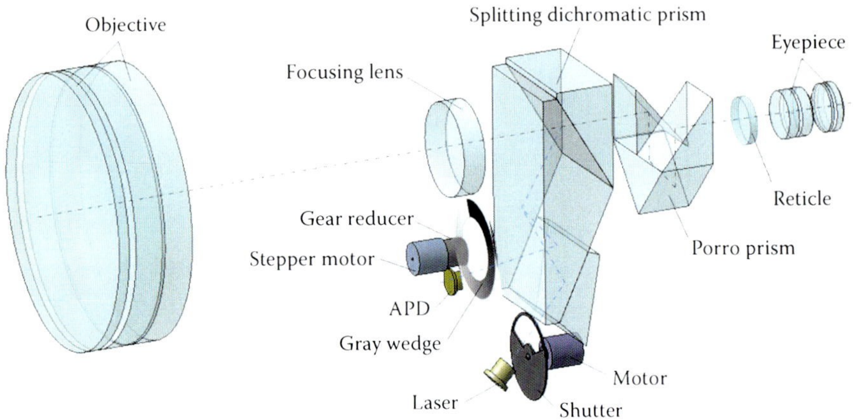

FIGURE 5.11 Telescope of Nikon total station with a reflectorless EDM.

The measurement and calibration channels are switched by a motor with a shutter on its axis. In to-object distance measurement mode the laser light goes through the hole in the shutter. In calibration mode the light is reflected from the shutter surface and directly enters the photodetector.

This telescope design is only applied in the total stations in Table 5.3.

At the present moment, manufacturers have slightly changed the optical scheme (Figure 5.13). The focusing lens is placed behind the splitting dichromatic prism. This allows for making the most compact telescope and the EDM distance range doubles. This type of a total station does not allow for shrinking the laser emission into a small spot on the object. Now the spot is about 30 mm and does not depend on the measured distance.

Nikon's new telescope forms a wide infrared light beam. The beam occupies almost the whole aperture of the objective and has little divergence. A red laser pointer is also built in the telescope. However, the laser pointer is not the part of the EDM. The red

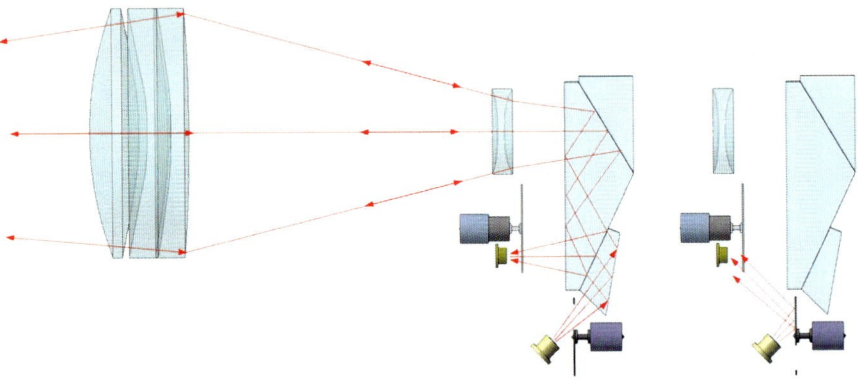

FIGURE 5.12 Ray tracing in EDM shown in Figure 5.11.

TABLE 5.3
Reflectorless Total Stations Provided with Nikon Telescope

Model	Distance to the Object/ Prism (m)	Accuracy: Distance Measurement (mm)	Accuracy: Angle Measurement (″)	Telescope	Manufacturer
NPL-332	200/2600	±(2 + 2 ppm × D)	5	Old	Nikon
NPL-352	200/2600	±(2 + 2 ppm × D)	2(Hz)/5(V)	Old	Nikon
NPL-322+	400/3000	±(3 + 2 ppm × D)	2(Hz)/5(V)	New	Nikon
Nivo C	400/3000	±(3 + 2 ppm × D)	1, 2, 3, 5	New	Nikon
Focus 6	400/3000	±(3 + 2 ppm × D)	2, 5	New	Spectraprecision
M3 DR	400/3000	±(3 + 2 ppm × D)	1, 2, 3, 5	New	Trimble

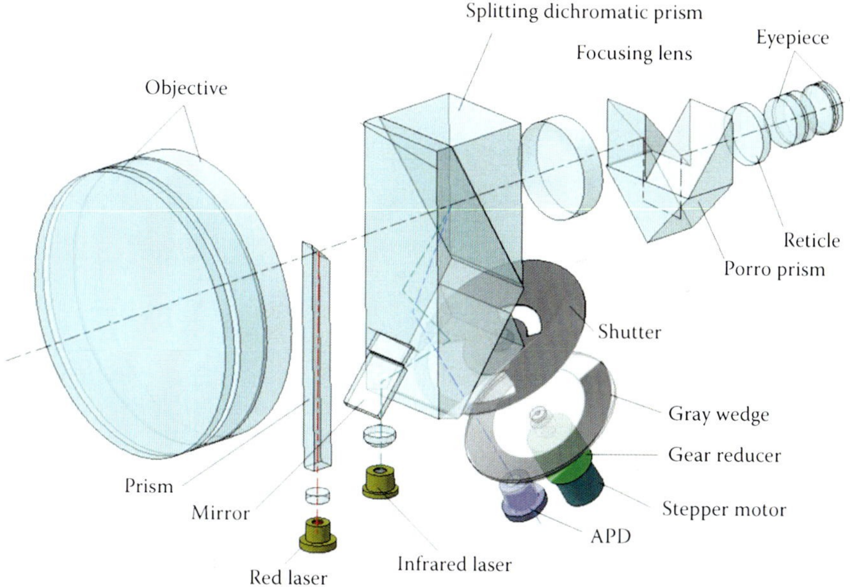

FIGURE 5.13 Nikon's new telescope.

light from the pointer is transmitted to the objective axis through the narrow prism (see Figure 5.13).

Some makers of total stations based on the Nikon optical design are listed in Table 5.3.

5.2 ADDITIONAL APPLIANCES OF A TOTAL STATION TELESCOPE

So far we have considered ordinary total station telescope optical design. Nowadays, telescopes may have additional features, including an automatic focusing system,

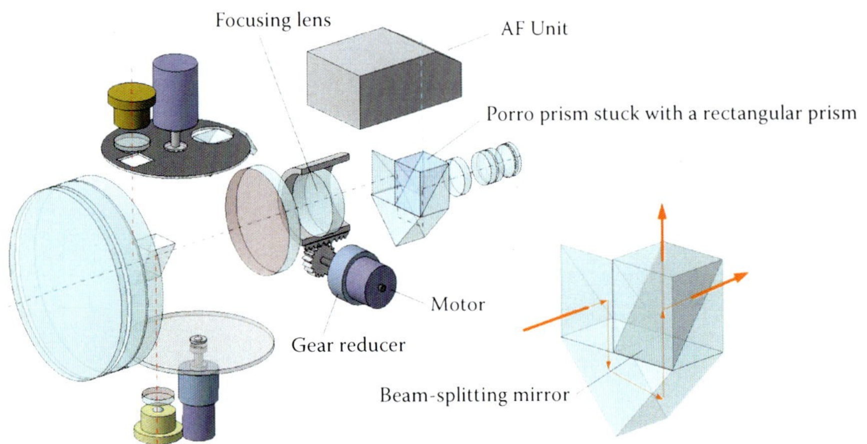

FIGURE 5.14 Telescope with an autofocus unit.

charge-coupled device (CCD) matrixes for photographing, a guide pointer, or an auto tracking system.

The autofocusing principle of operation is described in Chapter 2. An autofocusing system consists of an electronic and an optical part. The autofocusing electronic part is the same as autofocusing levels have. In Figure 5.14, we can see how the autofocus optical part is built into the reflectorless total station telescope. The focusing lens has a motor driver.

The Porro inverting prism together with a rectangular prism attached to one of its faces is used. The sticking site has a semitransparent mirror covering. One part of the beam is reflected from this mirror and runs to the ocular.

The other part of the beam goes through the semitransparent mirror and enters the sensor of autofocusing. A total station of this kind also has a manual focusing mode that is used for the battery energy saving. Also we use the manual mode when we work in imperfect illumination or if there are obstacles in the way such as leaves and branches of trees, or dense fencing. Nevertheless in good operating conditions, the autofocusing mode allows for time savings.

At first many Pentax models had autofocusing systems. Nowadays, only robotic total stations have autofocus.

Lately, all the leading companies produce total stations that can make video and photo of the object by means of digital CCD cameras built into their telescopes (Figure 5.15). In this case, the Porro prism attached to a rectangular prism is used to divide the sighting and video channels of the telescope.

A guide light unit appeared in total stations in the 1990s. The guide light is helpful for better layout. While doing layout the surveyor's assistant is able to see the reference line. The typical design is shown in Figure 5.16.

There is an additional objective above the total station main objective. The additional objective aperture is divided into two parts with a prism with mirror faces. Every part of the aperture is lighted with red or green LEDs. In case the LEDs are of the same color they blink in a different way.

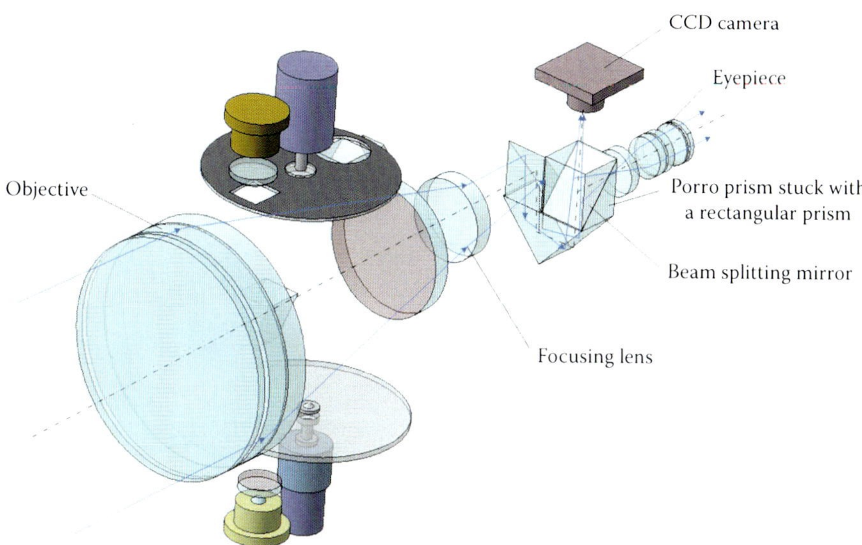

FIGURE 5.15 Telescope with a CCD camera.

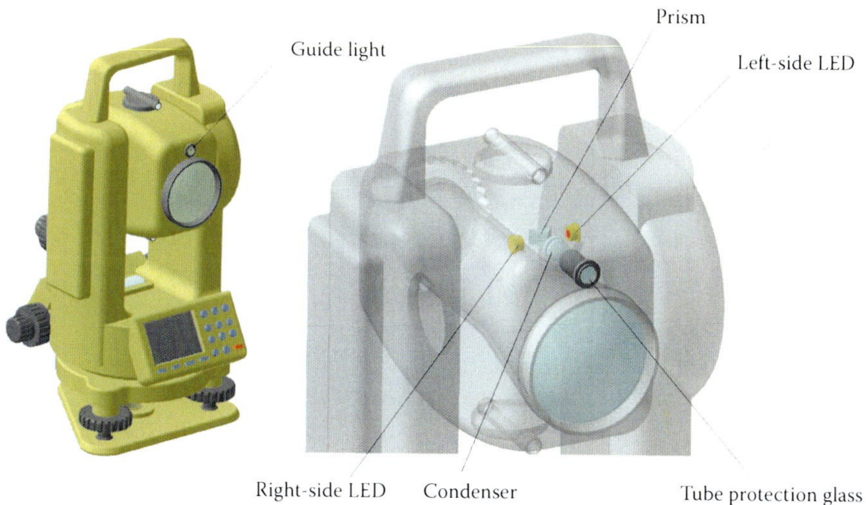

FIGURE 5.16 Telescope with a guide light.

5.3 ROBOTIC TOTAL STATIONS

Using a total station includes many routine activities. In order to automatize them, robotic total stations have appeared. They differ from ordinary total stations in movement of the horizontal and the vertical axes. Angular sensors and EDMs remained the same (Figure 5.17).

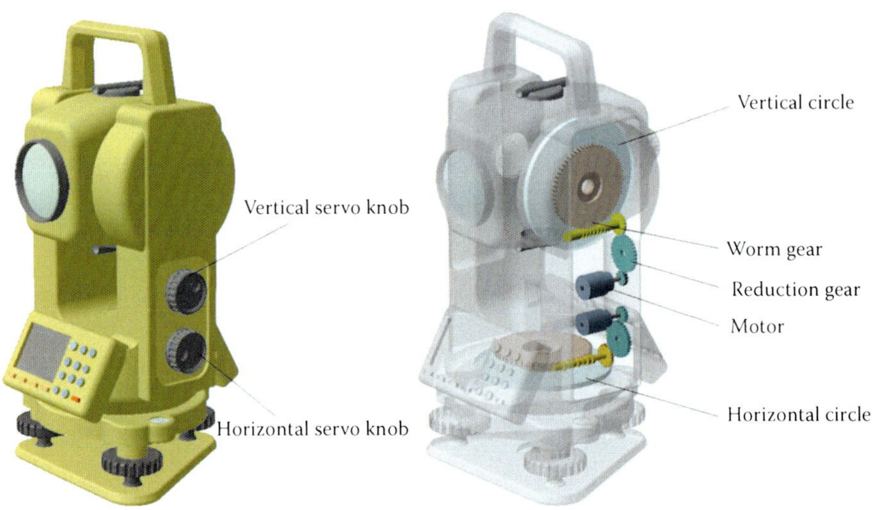

FIGURE 5.17 Robotic total station with reduction gear.

The total station axes are added with gears that are rotated by means of worm gear. The worm gear is moved by motor with a reduction gear. The total station computer controls these motors either with the help of the computer program or by servo control knobs. These knobs are placed on rotary encoders. The encoders are incremental sensors of angles. These sensor signals enter the computer, which processes them to control drivers. The program is made so that using the same knobs we can rotate the instrument slowly or quickly. When we rotate the knobs slowly, the instrument rotation is slow similar to using the theodolite tangent screws. When we rotate the knobs quickly, the instrument rotation speed exponentially increases, and with several turns of the knobs we can turn the instrument at a wide angle.

Lately the most advanced technologies of direct drivers are applied in some robotic total stations. The motor drivers are placed directly on the axes and gear reducers are not required.

Trimble offered an unconventional attractive solution of placing the servo motors directly on the total station rotation axes (Figure 5.18).

The servo motor is combined with the angular sensor into an integrated mechanical unit. The total station limbs have two tracks: barcode and incremental. The angular sensors enables reading the barcode and incremental tracks at the same time. The barcoded track on the limb is used for angle finding with a low accuracy. The incremental track is used for angle finding with a high accuracy. The final result of the measurements is obtained by computer processing of these two parts.

The servo motor is made under the patent technology MagDrive. The servo motor has two groups of powerful magnets placed in two rings. A powerful magnetic field occurs in the spacing between these two rings. In the magnetic field there is an annular frame with three windings coiled round along the frame. Current pulses in succession enter the windings under the controlling computer program. Depending on the program, the servo motors allow operation in one of three modes. The first mode

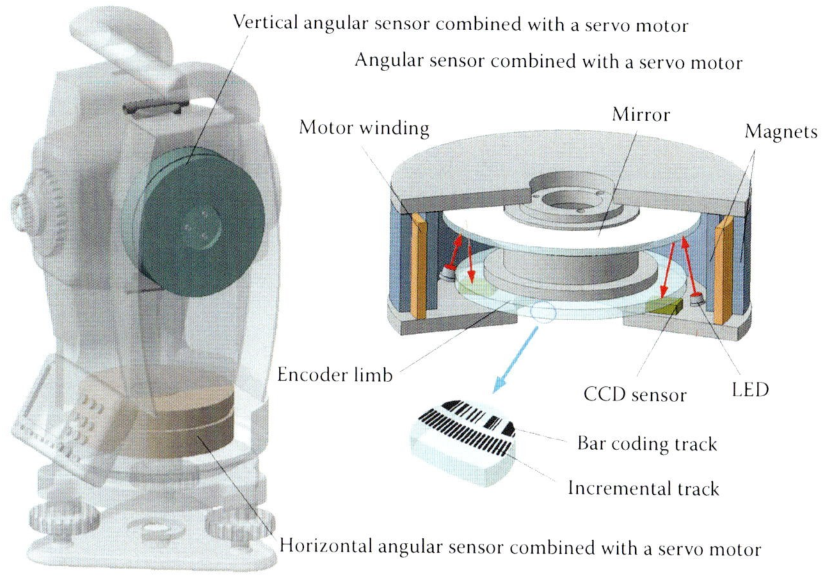

FIGURE 5.18 Robotic total station with servo motors.

is the magnet's rotation to one or another side relative to the frame. The second mode is servo motor rotation blocking. The third mode is servo motor deceleration. During deceleration the servomotors do not rotate the instrument, but there is opportunity to rotate the instrument by some hand force. Servo motors allow the total station to rotate two to three times quicker than in a classical scheme where reduction gears are applied. Servo motors are more reliable and efficient.

Another axes driver solution for a robotic total station is Leica's piezoelectric high-speed driver. The piezoelectric driver of the vertical axis is shown in Figure 5.19. The

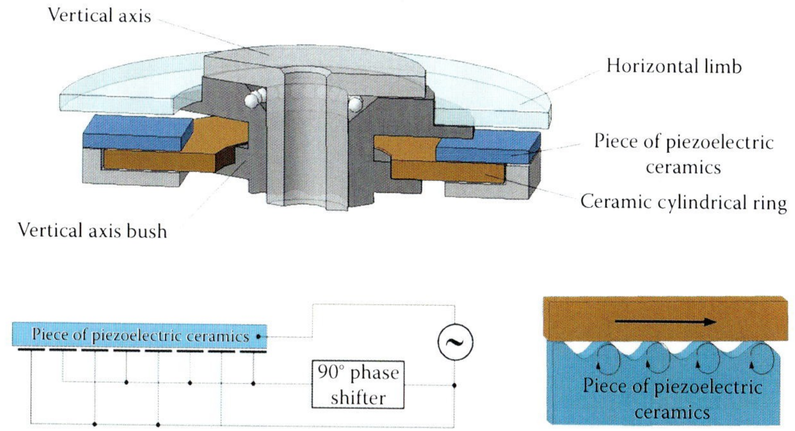

FIGURE 5.19 Vertical axis unit with a piezoelectric driver.

main part of the piezoelectric driver is piezoelectric ceramics. This kind of ceramics changes its linear size under electric field influence. Two alternate groups of electrodes are placed on a piezoelectric ceramic plate. When AC signals shifted at 90°, enter these groups the microwaves on the plate surface start moving. Once contact is made with the piezoelectric ceramic plate, the ceramic cylindrical ring moves. The movement direction and speed depends on phase shift, frequency, and the signals' amplitude. The piezoelectric driver is noiseless as ultrasonic signals are used in them.

As mentioned earlier, the robotic total station enables one to operate in autolock mode. In order to operate in this mode, we need a special circular reflector. Popular types of these reflectors are shown in Figure 5.20. They consist of several prism reflectors circularly placed.

FIGURE 5.20 Circular reflectors used in autolock.

Figure 5.21 illustrates a robotic total station telescope equipped with an autolock unit. Dichromatic prisms are used to combine the total station and autolock channels. Lasers with different wavelengths are used to divide the EDM and autolock channels. For example the following combination is possible: the EDM emission is 0.65 mkm and the autolock emission is 0.78 mkm.

The splitting dichromatic prism has two reflecting faces with different coverings. The front face has a narrow spectrum covering that reflects a 0.65 mkm wavelength. The diagonal face covering reflects light of only 0.78 mkm wavelength. The autolock unit has a CCD video matrix as a photodetector. In the past, a four-diode matrix was applied. If the laser spot gets into the center of the prism reflector, the spot must be in the center of the photodetector. If the spot is not in the center of the photodetector, the total station computer estimates the miscoordination and sends signals to

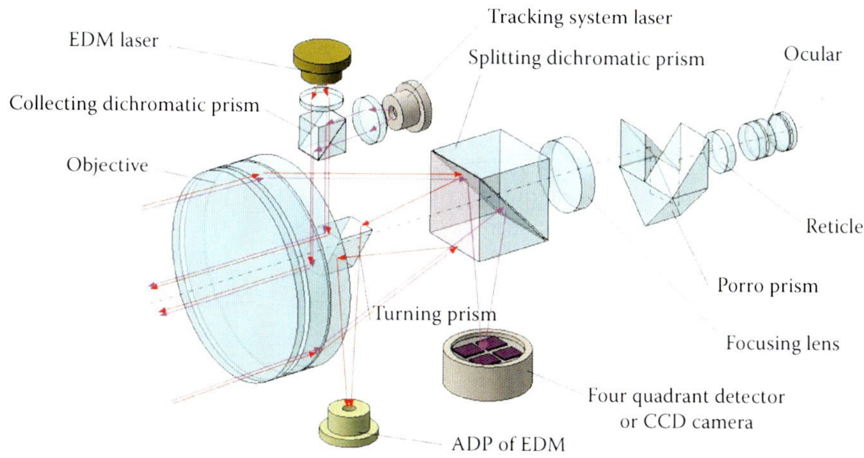

FIGURE 5.21 Robotic total station telescope with a reflectorless EDM and autolock unit.

corresponding drivers of the instrument's axes. The drivers rotate the instrument until the light spot appears exactly in the middle of the photodetector.

Table 5.4 shows the leading manufacturers of robotic total stations.

5.4 CHECKING AND ADJUSTING TOTAL STATIONS

At the beginning of this chapter we paid attention to the fact that a total station consists of a theodolite and EDM. Thus let us start checking the theodolite parts. For that we must verify the following parameters:

1. Collimation error
2. Vertical index
3. Telescope rotation axis inclination
4. Perpendicularity of the tubular/electronic level and the total station rotation axis
5. Coincidence of the total station rotation axis with the optical/laser plummet
6. Parallelity of the circular level and the total station rotation axis

The theodolite part checking and adjustment in no way differs from an electronic theodolite inspection. A total station is provided with a smarter user's menu, for it has a more powerful computer than an electronic theodolite. The menu contains a special item—"Calibration"—which allows carrying out some adjustment. Thus we can correct the collimation error, the vertical index, and the electronic level. High-precision total stations have an additional item in the menu that allows taking into account inclination of the telescope rotation axis.

Now let us discuss the problems concerning the EDM part of a total station. Before starting to check and adjust the EDM we should be sure of the operating parameters' compliance in the theodolite part of the total station. We must pay much attention to the collimation error value. It is to be minimal or at least permissible.

TABLE 5.4
Up-to-Date Robotic Total Stations

Model	Angle Measurement Accuracy (″)	Distance Measurement Accuracy (mm) Prism/Reflectorless	Max. Measurement Distance: Reflectorless/Single Prism (km)	Drives	Interface	Manufacturer
TS15G R400	1, 2, 3, 5	±(1 + 1.5 ppm × D)/ ±(2 + 2 ppm × D)	0.4/10	Servo motor	RS-232 Bluetooth USB	Leica Geosystems
TS15G R1000	1, 2, 3, 5	±(1.5 + 2 ppm × D)/ ±(2 ÷ 4 + 2 ppm × D)	1.0/10	Servo motor	RS-232 Bluetooth USB	Leica Geosystems
TS16 R500	1, 2, 3, 5	±(1.5 + 2 ppm × D)/ ±(2 + 2 ppm × D)	0.5/10	Servo motor	RS-232 Bluetooth USB	Leica Geosystems
TS16 R1000	1, 2, 3, 5	±(1.5 + 2 ppm × D)/ ±(2 ÷ 4 + 2 ppm × D)	1.0/10	Servo motor	RS-232 Bluetooth USB	Leica Geosystems
MS60	1	±(1.5 + 2 ppm × D)/ ±(2 ÷ 4 + 2 ppm × D)	2.0/10	Piezo motor	RS-232 Bluetooth USB	Leica Geosystems
MS60 I	0.5	±(0.6 + 1 ppm × D)/ ±(2 ÷ 4 + 2 ppm × D)	1.0/10	Piezo motor	RS-232 Bluetooth USB	Leica Geosystems
TM50	0.5, 1	±(0.6 + 1 ppm × D)/ ±(2 + 2 ppm × D)	1.0/3.5	Piezo motor	RS-232 Bluetooth USB	Leica Geosystems
SX-100	1, 3, 5	±(1.5 + 2 ppm × D)/ ±(2 + 2 ppm × D)	1.0/6	Servo motor	RS-232 Bluetooth USB	Sokkia
S5	1, 2, 3, 5	±(1 + 2 ppm × D)/ ±(2 + 2 ppm × D)	1.3/2.5	MagDrive	RS-232 USB	Trimble
S7	1, 2, 3, 5	± (1 + 2 ppm × D)/± (2 + 2 ppm × D)	1.3/2.5	MagDrive	RS-232 USB	Trimble
S9HP	0.5, 1	±(0.8 + 2 ppm × D)/ ±(3 + 2 ppm × D)	0.15/3	MagDrive	RS-232 USB	Trimble
VX10	1	±(1 + 1.5 ppm × D)/ ±(2 + 1.5 ppm × D)	0.8/5.5	MagDrive	RS-232 USB	Trimble

A total station EDM should meet the following requirements:

1. The EDM axis should coincide with the telescope collimation axis.
2. Distance measurement accuracy should correspond to total station technical data.

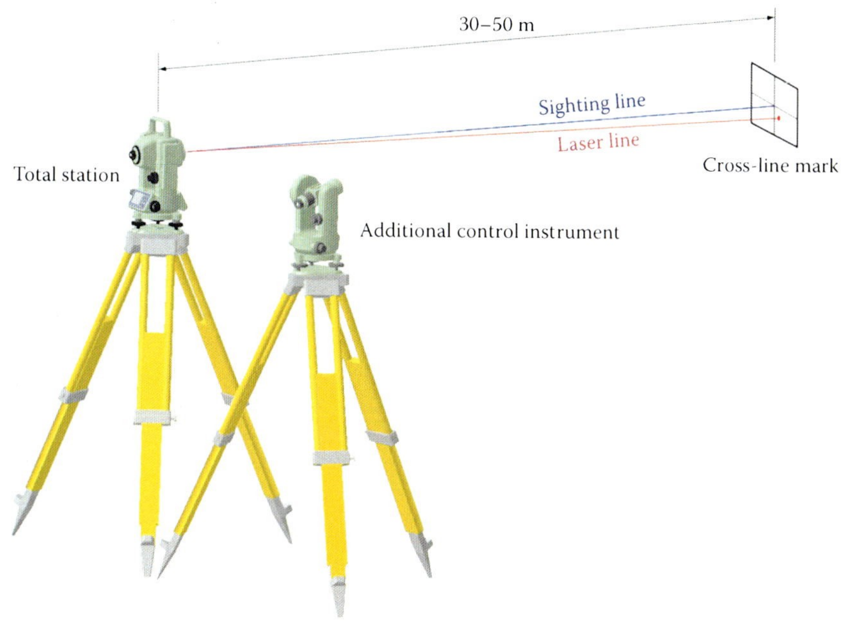

FIGURE 5.22 EDM axis checking.

Let us start from the first requirement and discuss what an EDM axis is. We recall that the telescope collimation axis is a straight or broken line going through the objective center and the reticle crossing. Similarly we can say this about the EDM axis. More specifically an EDM has two axes. These are the transmitting and receiving channels axes. The EDM transmitting channel axis is a line (straight or broken) going through the objective center and the EDM laser. Similarly the receiving channel axis connects the objective and the photodetector centers.

We can check the telescope optical and the transmitting channel axes coincidence. The receiving channel axis can be checked only at a workshop.

If the total station EDM has a red laser, we can easily check the theodolite and the EDM axes coincidence (Figure 5.22). Let us put the total station on the tripod and then place a cross-line mark on the wall approximately at the same level with the instrument at the distance of 30–50 m. If we have a theodolite at hand we put it next to the total station to watch the mark. In case there is no theodolite, we need an assistant who can watch the spot position on the mark. The total station provided with a visual laser cannot be applied as an auxiliary instrument, since its splitting dichromatic prism in the telescope does not allow watching a laser spot on the mark.

Let us switch the total station into laser pointer mode and superpose the reticle crossing with the center of the mark. In case the total station does not have this mode, we can run the EDM tracking mode. Ideally the laser spot is to enter exactly the center of the mark. If it does not, we have to measure the deviation with the total station angular system. For that we superpose the laser spot with the center of the

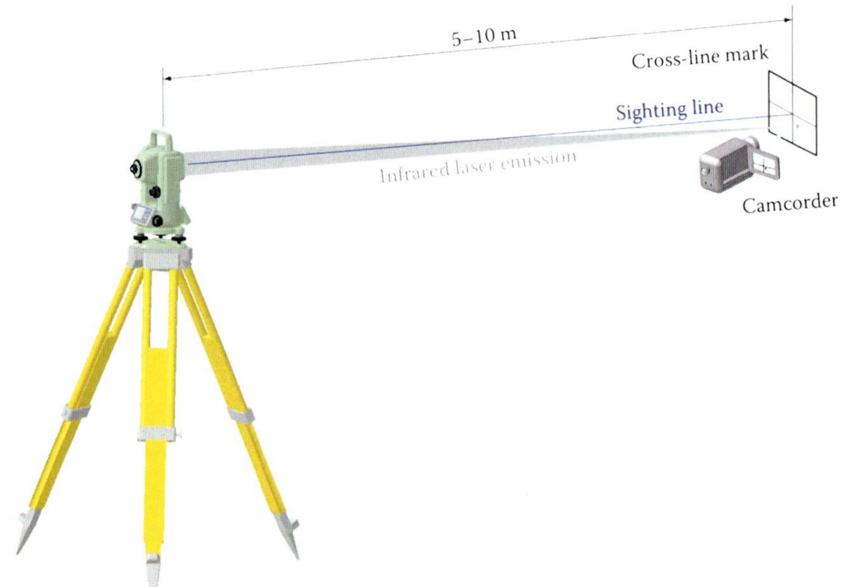

FIGURE 5.23 Infrared EDM axis checking.

mark and read out the angular deviation. If the deviation is not significant (within 30″) we have a chance to eliminate it by ourselves. Pentax total stations and some Trimble total stations have two holes closed with rubber corks in their telescope covers. Under them there are regulation screws that are used to adjust the laser line direction. In case these screws are not provided, little disparities of the theodolite and the EDM axes may have been eliminated by the reticle correction screws.

Usually we do not have trouble with the vertical adjustment of the laser beam. If we vertically superpose the laser line with the theodolite optical axis by the means of the reticle correcting screws, the vertical index value changes. In this case it may be easily corrected with the help of routine adjusting of the total station.

The laser beam horizontal adjustment may cause trouble due to the collimation error value change as a result the reticle screws rotation. The collimation error can be eliminated using the point in the "Calibration" menu option of the total station. However, the error may be eliminated only within 20″.

Some reflectorless total stations EDMs have an infrared laser. In order to watch a laser spot on the mark we can use a camcorder with night vision (Figure 5.23).

We can check the coincidence between EDM and collimation axes another way. This way is applicable in reflectorless EDMs with any laser type. Let us turn the mark to approximately 45° relative to the telescope axis. When the instrument is in position I we superpose the crosshairs with the mark center and measure the distance for the first time. Then we turn the telescope to position II and measure the distance again. The coincidence of the axes is acceptable if the two measurements results do not differ more than the EDM double standard deviation. As well the test should be done in the position of the mark vertical inclination.

The instrument constant should be checked from time to time but not less than twice a year, especially after impact and drops. Instrument constant checking is necessary both in reflectorless and in reflector modes.

Standard deviation is used to provide accuracy of distance measurement. In the technical data of a total station it looks like this:

$$\sigma mm = Amm + Bppm \tag{5.1}$$

where *Amm* is the constant component of the distance measurement error not depending on the distance, specified in millimeters; and *Bppm* is the variable component of distance measurement error depending on the distance, specified as the number of millimeters per every kilometer of distance. One ppm equals 1 mm/km.

We can measure distance with the accuracy referred to in the technical data only if all possible user's mistakes are eliminated. The most common users' mistakes influencing upon *Amm* are

- Incorrect input of the instrument constant in the total station menu.
- Incorrect input of the prism constant in the total station menu. This mistake occurs only while measuring distance in a prism mode. The mistake may sometimes occur when setting different prisms constant or in case of the wrong constant sign.

As usual, the prism constant value is written on the frame where the reflector is set up. Sometimes the minus sign is not present in the prism constant value. Also in total stations made by different manufacturers, the prism constant sign is interpreted in different ways. If in doubt about the sign, we advise to measure a check line even with a low accuracy. For that, using a tape measure we can lay out a 10–20 m line on the horizontal surface and then measure it with the total station. The prism constant sign is immediately found out. If the total station operates in reflectorless mode, we can do it without a measuring tape. Since the prism reflector is provided with a target plate for angular measurement, we can use it to find out the prism constant value.

If the instrument is dropped or impacted, *Amm* may have been changed because the EDM optical parts shifted. Also, decreasing of the total station distance range may occur. In this case, the instrument should be checked at a service workshop.

The most common users' mistakes influencing *Bppm* are

- Input of incorrect scale factor (another name is grid factor) in the total station menu.
- Software of some total stations enables changing the scale factor. After using this option it is necessary to put the scale factor back to 1.000000.
- Input of incorrect values for both air temperature and pressure in the total station menu. If the temperature value differs from the real one at 10°C the scale factor will change at about 10 ppm (or 1 mm per every 100 m). The same error will occur if the atmospheric pressure value does not corrected after moving at a height of 350 m above sea level.

Sometimes in EDMs, the quartz oscillator frequency referential meaning changes. It can be discovered after measuring several referential distances. In this case *Bppm* can be corrected only by regulating the quartz oscillator frequency referential meaning at a service workshop.

Unassisted EDM constant precise checking is complicated enough. Only in specialized workshops or metrological offices is there referential meaning distances for checking.

We cannot recommend ways to find the instrument constant by measuring the line as a whole and then in parts. This method is laborious enough and does not provide accuracy, since the final result will include the mistakes of all three distance measurements. Also, an incorrect scale factor input is not discovered when applying this method.

The method of direct collation using a higher precision total station or several identical ones is more reliable. The measurements should be carried out in minimal atmospheric refractions. The tripod with a tested total station is put on a stable plane surface. The checking is carried out both in reflectorless and reflector modes. In order to fulfill checking in reflectorless mode we lay out three lines at about 5, 30, and 150 m. Then we need to fulfill checking in reflector mode and set up two prisms (reflectors). They are set up at the distances of ten and several hundred meters, respectively.

Then, we measure these five lines several times and average the result. Then, we carefully disconnect the total station from the tripod and put onto its place a reference one. Then, using the reference total station we measure the same lines and compare the measurements results. If the measurement results differ from the standard deviation difference twice, the instrument constant value should be corrected. Unfortunately, the instrument constant correction is not available in the menu of every total station. If the difference is considerable the instrument should be delivered to a service workshop.

BIBLIOGRAPHY

Gächter, B. and B. Braunecker. 2009. Geodesic device comprising laser source. US Patent 7,480,316 B2 filed September 17, 2004, and issued January 20, 2009.

Herbst, C., H. Bernhard, and A. Häle. 2011. Geodesic instrument with a piezo drive. US Patent 7,999,921 B2 filed July 22, 2005, and issued August 16, 2011.

Johnson, J. and A. Utterback. 2009. Surveying instrument and method of controlling surveying instrument. US Patent 7,640,068 filed June 25, 2007, and issued December 29, 2009.

Katsuma, M., H. Murakami, and A. Hiramatsu. 1985. Vibration wave motor. US Patent 4,513,219 filed November 16, 1983, and issued April 23, 1985.

Kimura, A. 1995. Rotating and driving system for survey instrument. US Patent 5,475,930 filed June 27, 1994, and issued December 19, 1995.

Kludas, T., A. Schünemann, and M. Vogel. 2011. Surveying instrument and method of providing survey data of a target region using surveying instrument. US Patent 8,024,144 B2 filed September 11, 2006, and issued September 20, 2011.

Kludas, T. and M. Vogel. 2010. Surveying method and surveying instrument. US Patent 7,830,501 B2 filed October 26, 2005, and issued November 9, 2010.

Niiho, M., K. Itoh, S. Suzuki, T. Itagaki, and K. Tsuda. 1987. Change-over shutter for light-wave range finger. US Patent 4,636,068 filed April 18, 1983, and issued January 13, 1987.

Ohishi, M. 2001. Distance measuring system. US Patent 6,333,783 B1 filed April 26, 2000, and issued December 25, 2001.

Ohishi, M. and Y. Tokuda. 2002. Distance measuring apparatus. US Patent 6,384,904 B1 filed May 19, 2000, and issued May 7, 2002.

Shrai, M. 2002. Surveying instrument having an optical distance meter. US Patent 6,469,777 B2 filed June 11, 2001, and issued October 22, 2002.

Tanaka, T. 2001. Distance measuring apparatus. US Patent 6,252,655 B1 filed July 7, 1998, and issued June 26, 2001.

Westermark, M., M. Hertzman, U. Berg, and T. Klang. 2010. Position control arrangement, especially for a surveying instrument, and a surveying instrument. US Patent 7,765,084 B2 filed November 10, 2008, and issued July 27, 2010.

6 Global Navigation Satellite Systems (GNSS)

Global Navigation Satellite Systems (GNSS) include constellations of Earth-orbiting satellites that broadcast their locations in space and time, of networks of ground control stations, and of receivers that calculate ground positions by trilateration.

"Education Curriculum: Global Navigation Satellite Systems," United Nations Office for Outer Space Affairs

6.1 FUNDAMENTAL PRINCIPLES OF GLOBAL NAVIGATION SATELLITE SYSTEMS (GNSS)

6.1.1 Principle of Signal Transit Time Measuring

At some time or another during a thunderstorm, you have almost certainly attempted to figure out how far away you are from a bolt of lightning. The distance can be established quite easily (see Figure 6.1): The distance is equal to the time a lightning flash is perceived (start time) up until the thunder is heard (stop time) multiplied by the speed of sound (approximately 330 m/s). The difference between the start and end time is referred to as the signal travel time. In this case, the signal is sound waves traveling through the air.

$$\text{Distance} = \text{Travel time} \times \text{Speed of sound}$$

Satellite navigation functions based on the same principle. One calculates the position by establishing the distance relative to reference satellites with a known position. In this case, the distance is based on the travel time of radio waves transmitted from the satellites.

6.1.2 Basic Principles of Satellite Navigation

Global Navigation Satellite Systems (GNSS) all use the same basic principles to determine coordinates:

- Satellites with a known position transmit a regular time signal.
- Based on the measured travel time of the radio waves (electromagnetic signals travel through space at the speed of light $c = 300{,}000$ km/s), the position of the receiver is calculated.

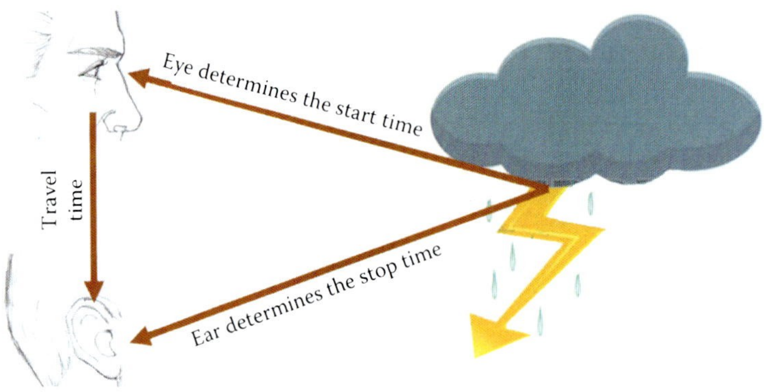

FIGURE 6.1 Determining the distance of a lightning flash.

We can see the principle more clearly using a simple model. Imagine that we are in a car and we need to determine our position on a long and straight street. At the end of the street there is a radio transmitter sending a time signal pulse every second. Onboard the car we are carrying a clock synchronized to the clock at the transmitter. By measuring the elapsed travel time from the transmitter to the car we can calculate our position on the street (Figure 6.2).

The distance D is calculated by multiplying the travel time $\Delta\tau$ by the velocity of light c:

$$D = \Delta\tau * c \tag{6.1}$$

Because the time of the clock onboard of our car may not be exactly synchronized with the clock at the transmitter, there can be discrepancies between the calculated and actual distance traveled. In navigation, this observed distance referenced to the local clock is referred to as pseudorange. In our example, a travel time $\Delta\tau$ of one microsecond (1 μs) generates a pseudorange of 300 m.

We could solve the problem of local clock synchronization by equipping our car with an exact atomic clock, but this would probably exceed our budget. Another solution

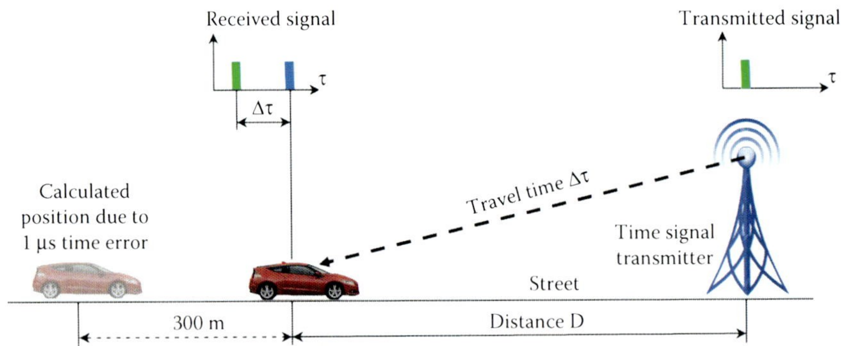

FIGURE 6.2 In the simplest case, the distance is determined by measuring the travel time.

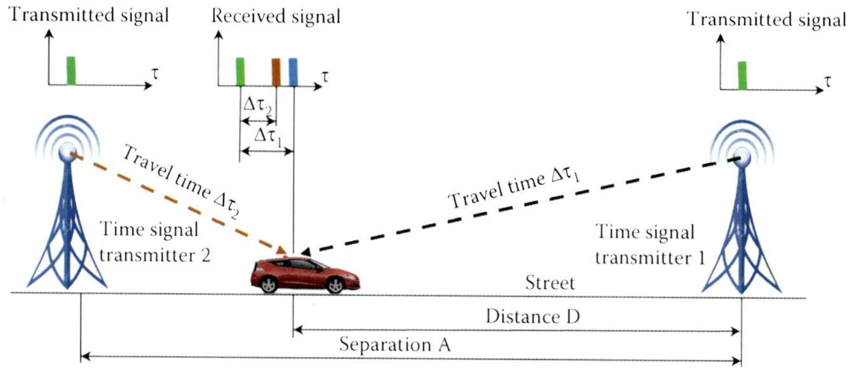

FIGURE 6.3 With two transmitters, it becomes possible to calculate the exact position despite time errors.

involves using a second synchronized time signal transmitter, for which the separation (A) to the first transmitter is known. By measuring both travel times, it is possible to establish the exact distance (D) despite having an imprecise onboard clock (Figure 6.3):

$$D = \frac{(\Delta\tau_1 - \Delta\tau_2) * c + A}{2} \tag{6.2}$$

As we have seen, in order to calculate the exact position and time along a line (by definition a line expands in one dimension) we require two time signal transmitters. From this, we can draw the following conclusion: When employing an unsynchronized onboard clock for the purposes of calculating the position, it is necessary that the number of time signal transmitters exceed the number of unknown dimensions by a value of one. For example:

- On a plane (expansion in two dimensions) we need three time-signal transmitters
- In three-dimensional space we need four time-signal transmitters

Satellite Navigation Systems use satellites as time-signal transmitters. Connection to at least four satellites (see Figure 6.4) is necessary in order to determine the three desired coordinates (longitude, latitude, and altitude) as well as the exact time.

6.1.3 Signal Travel Time

Satellite Navigation Systems employ satellites orbiting high above the Earth and which are spread out in such a way that there is line-of-sight connections to at least four satellites from any point on the ground. Each one of these satellites is equipped with onboard atomic clocks. Atomic clocks are the most precise time measurement instruments known, losing a maximum of one second every 30,000 to 1,000,000 years. In order to make them even more accurate, they are regularly adjusted or

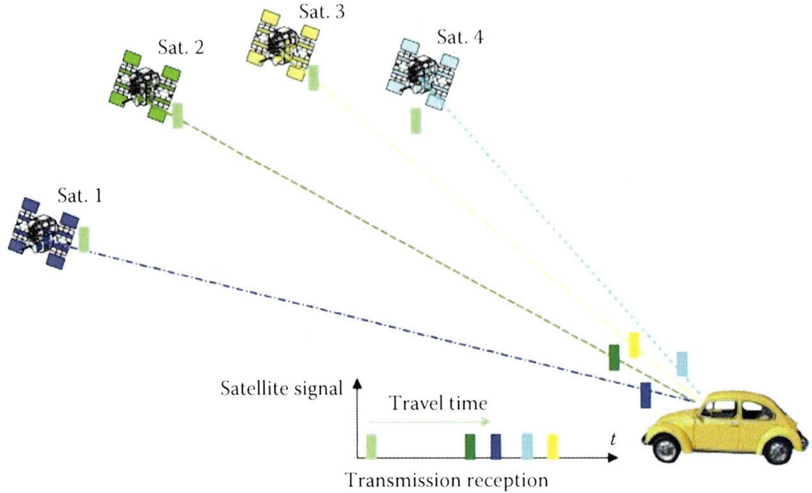

FIGURE 6.4 Four satellites are needed to determine longitude, latitude, altitude, and time.

synchronized from various control points on Earth. GNSS satellites transmit their exact position and onboard clock time to Earth. These signals are transmitted at the speed of light (300,000 km/s) and therefore require approximately 67.3 ms to reach a position on Earth's surface directly below the satellite. The signals require another 3.33 s for each additional kilometer of travel. To establish the position, all that is required is a receiver and an accurate clock. By comparing the satellite's signal arrival time with the onboard clock at the moment of signal transmission, it is possible to determine the signal travel time (Figure 6.5).

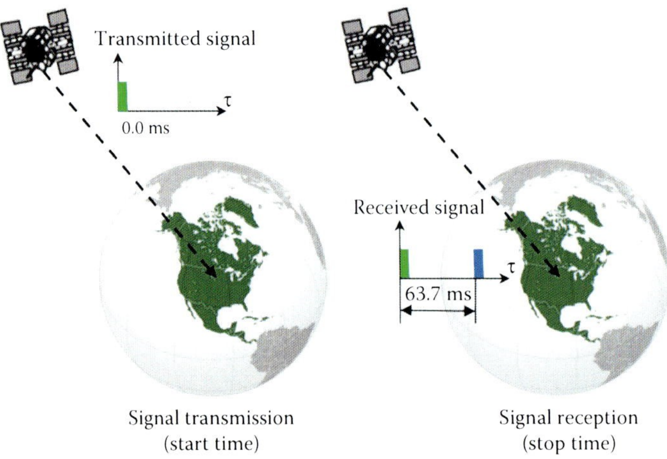

FIGURE 6.5 Determining the signal travel time.

As with the example of the car, the distance D to the satellite can be determined from the known signal travel time $\Delta\tau$:

$$\text{Distance} = \text{Travel time} \times \text{Speed of sound}$$

$$D = \Delta\tau * c$$

6.1.4 Determining the Position

Imagine that you are wandering across a vast plateau and would like to know where you are. Two satellites are orbiting far above you transmitting their onboard clock times and positions. By using the signal travel time to both satellites, you can draw two circles with the radii D_1 and D_2 around the satellites. Each radius corresponds to the calculated distance to the satellite. All possible positions relative to the satellites are located on these circles. If the position above the satellites is excluded, the location of the receiver is at the exact point where the two circles intersect beneath the satellites (Figure 6.6), therefore, two satellites are sufficient to determine a position on the *X*/*Y* plane.

In the real world, a position has to be determined in three-dimensional space rather than on a plane surface. As the difference between a plane surface and three-dimensional space consists of an extra dimension (height *Z*), an additional third satellite must be available to determine the true position. If the distance to three satellites is known, all possible positions are located on the surface of three spheres whose radii correspond to the distance calculated. The position is the point where all of the three spheres intersect (Figure 6.7).

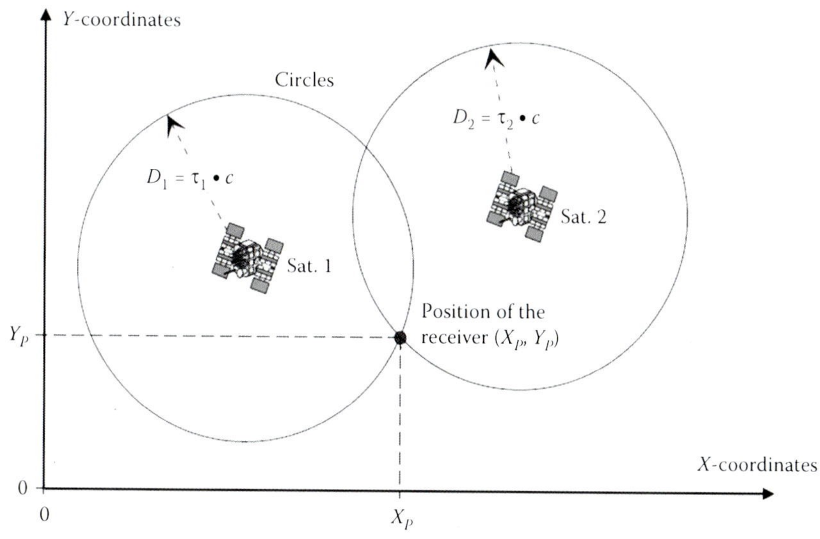

FIGURE 6.6 The position of the receiver at the intersection of the two circles.

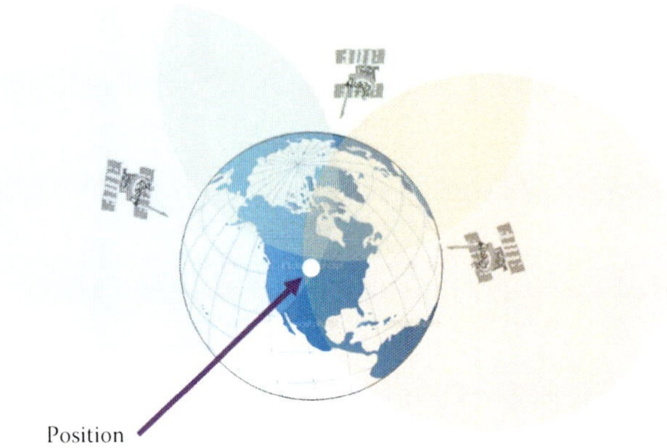

FIGURE 6.7 The position is determined at the point where all three spheres intersect.

6.1.5 Effect and Correction of Time Error

The conclusions in the previous section are valid only if the clock at the receiver and the atomic clocks onboard the satellites are synchronized, that is, the signal travel time can be precisely determined. If the measured travel time between the satellites and an earthbound navigational receiver is incorrect by just 1 μs, a 300 m position error of is produced. As the clocks onboard all the GNSS satellites are synchronized, the signal travel time in the case of all three measurements is inaccurate by the same amount. Mathematics can help us in this situation.

When performing mathematical calculations, we remember that if N variables are unknown, we need N independent equations to identify them. If the time measurement is accompanied by a constant unknown error (Δt), we will have four unknown variables in three-dimensional space:

- Longitude (X)
- Latitude (Y)
- Height (Z)
- Time error (Δt)

These four variables require four equations, which can be derived from four separate satellites.

Satellite Navigation Systems are deliberately constructed in such a way that from any point on Earth, at least four satellites are "visible" (Figure 6.8). Thus, despite an inaccuracy on the part of the receiver clock and resulting time errors, a position can be calculated to within an accuracy of approximately 5–10 m.

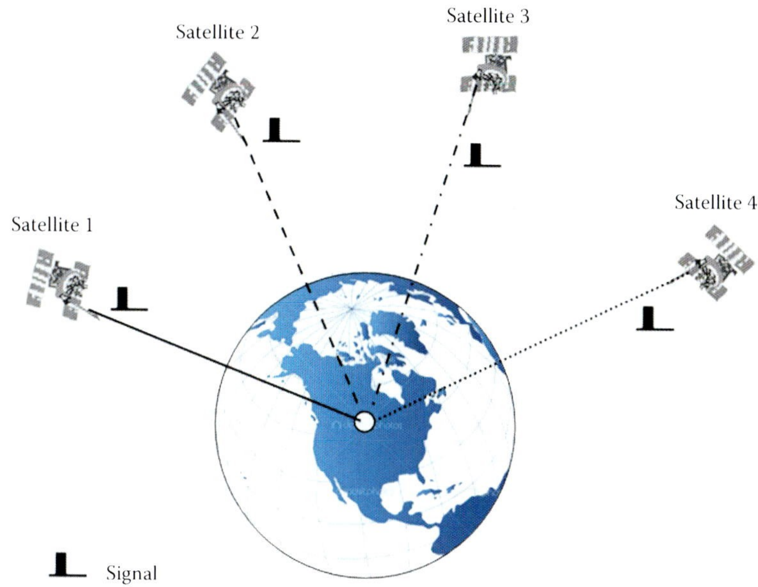

FIGURE 6.8 Four satellites are required to determine a position in 3D space.

6.2 COORDINATE SYSTEMS

A significant problem to overcome when using a GNSS is the fact that there are great numbers of different coordinate systems worldwide. As a result, the position measured and calculated does not always correspond with one's supposed position.

In order to understand how GNSS function, it is necessary to examine some of the basics of geodesy, the science that deals with the surveying and mapping of Earth's surface. Without this basic knowledge, it is difficult to understand the apparently bewildering necessity of combining the appropriate map reference systems (datums) and grids. Of these, there are more than 100 different datums and approximately 10 different grids to select from. If an incorrect combination is made, a position can be off by several hundred meters.

6.2.1 Geoid

We have known that the Earth is round since Columbus. But how round is it really? Describing the shape of our blue planet has always been a challenging scientific task. Over the centuries, several different models have been presented to represent an approximation of the true shape of the earth as accurate as possible.

The geoid represents the true shape of the earth, defined as the surface where the mean sea level is zero. This shape is defined by the gravity of the earth, thus its geometrical description is rather complex. Using the Greek word for Earth, this geometrical shape of this surface is called geoid (Figure 6.9).

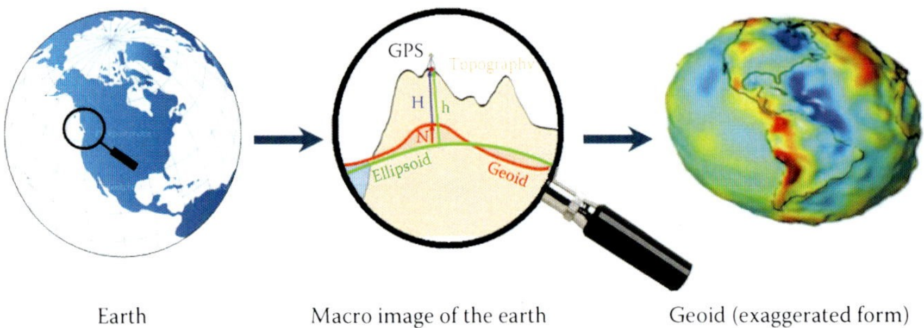

FIGURE 6.9 A geoid is an approximation of the Earth's surface.

Because the distribution of the mass of the Earth is uneven and, as a result, the level surface of the oceans and seas do not lie on the surface of a geometrically definable shape, approximations like ellipsoids have to be used. Differing from the actual shape of the Earth, a geoid is a theoretical body whose surface intersects the gravitational field lines everywhere at right angles.

A geoid is often used as a reference level for measuring height. For example, the reference point in Switzerland for measuring height is the "Repère Pierre du Niton" (RPN, 373.600 m) in the Geneva harbor basin. This height originates from point to point measurements with the port of Marseilles (mean height above sea level 0.00 m).

6.2.2 Ellipsoid and Datum

6.2.2.1 Ellipsoid

A geoid is a difficult shape to manipulate when conducting calculations. Therefore, a simpler, more definable shape is needed when carrying out daily surveying operations. Such a substitute surface is known as an ellipsoid. If the surface of an ellipse is rotated about its symmetrical North–South Pole axis, a spheroid is obtained as a result (Figure 6.10).

An ellipsoid is defined by two parameters:

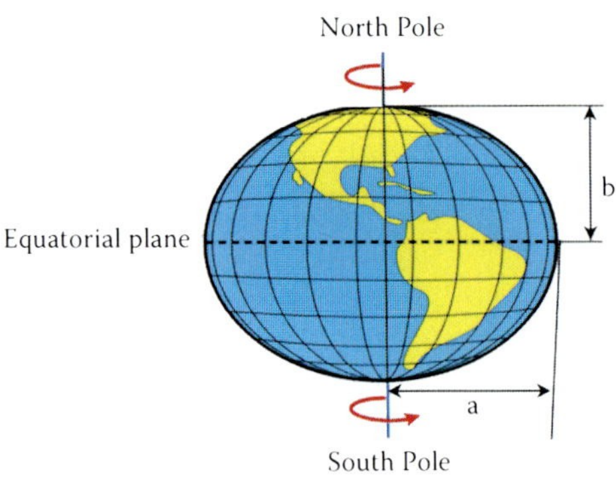

FIGURE 6.10 Producing a spheroid.

- Semimajor axis a (on the equatorial plane)
- Semiminor axis b (on the North–South Pole axis)

The amount by which the shape deviates from the ideal sphere is referred to as flattening (f):

$$f = \frac{a-b}{a} \tag{6.3}$$

6.2.2.2 Customized Local Reference Ellipsoids and Datum

6.2.2.2.1 Local Reference Ellipsoids

When dealing with an ellipsoid, one should be careful to ensure that the natural perpendicular does not intersect vertically at a point with the ellipsoid, but rather with the geoid. Normal ellipsoidal and natural perpendiculars do not therefore coincide; they are distinguished by "vertical deflection" (Figure 6.12), that is, points on the Earth's surface are incorrectly projected. In order to keep this deviation to a minimum, each country has developed its own customized nongeocentric ellipsoid as a reference surface for carrying out surveying operations (Figure 6.11). The semiaxes a and b as well as the midpoint are selected in such a way that the geoid and ellipsoid match national territories as accurately as possible.

6.2.2.2.2 Datums (or Map Reference Systems)

National or international map reference systems based on certain types of ellipsoids are called datums. Depending on the map used when navigating with GNSS receivers, you must be attentive to ensure that the relevant map reference system has been entered into the receiver.

There are over 120 map reference systems available, such as CH-1903 for Switzerland, NAD83 for North America, and WGS-84 as the global standard.

An ellipsoid is well suited for describing the positional coordinates of a point in degrees of longitude and latitude. Information on height is either based on the geoid or the reference ellipsoid. The difference between the measured orthometric height H, that is, based on the geoid, and the ellipsoidal height h, based on the reference ellipsoid, is known as geoid undulation N (Figure 6.12).

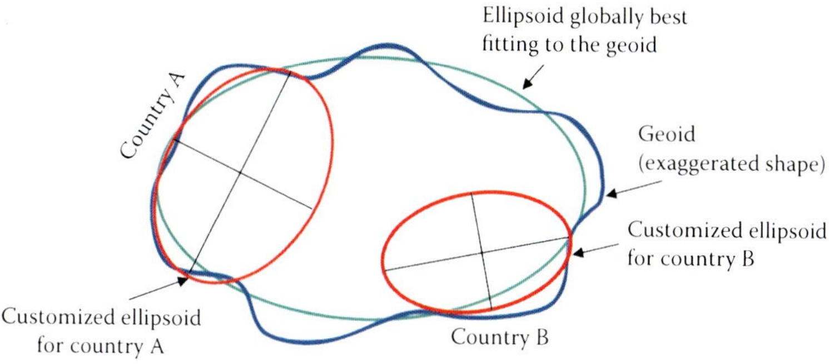

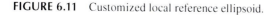

FIGURE 6.11 Customized local reference ellipsoid.

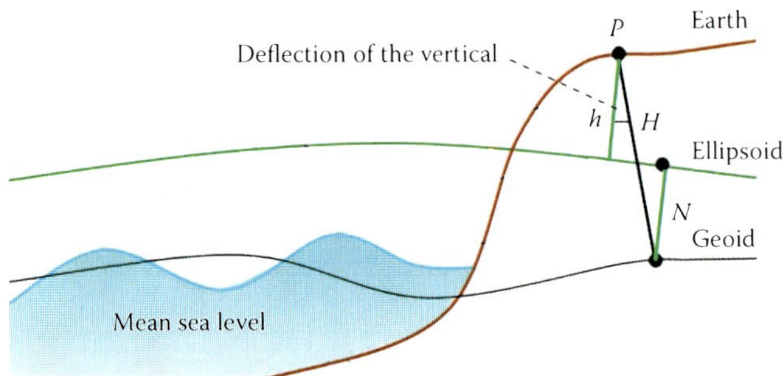

FIGURE 6.12 Difference between geoid and ellipsoid.

6.2.2.3 Worldwide Reference Ellipsoid WGS-84

The details displayed and calculations made by a GNSS receiver primarily involve the WGS-84 (World Geodetic System 1984) reference system. The WGS-84 coordinate system is geocentrically positioned with respect to the center of the Earth. Such a system is called ECEF (Earth centered, Earth fixed). The WGS-84 coordinate system is a three-dimensional, right-handed, Cartesian coordinate system with its original coordinate point at the center of the mass (geocentric) of an ellipsoid, which approximates the total mass of the Earth.

The positive *X*-axis of the ellipsoid (Figure 6.13) lies on the equatorial plane (that imaginary surface which is encompassed by the equator) and extends from the center of mass through the point at which the equator and the Greenwich meridian intersect (the 0 meridian). The *Y*-axis also lies on the equatorial plane and is offset 90° to the east of the *X*-axis. The *Z*-axis lies perpendicular to the *X*- axis and *Y*-axis and extends through the geographical North Pole.

The parameters of the WGS-84 ellipsoid are summarized in Table 6.1.

Ellipsoidal coordinates (φ, λ, h), rather than Cartesian coordinates (X, Y, Z) are generally used for further processing (Figure 6.14). φ corresponds to latitude, λ to

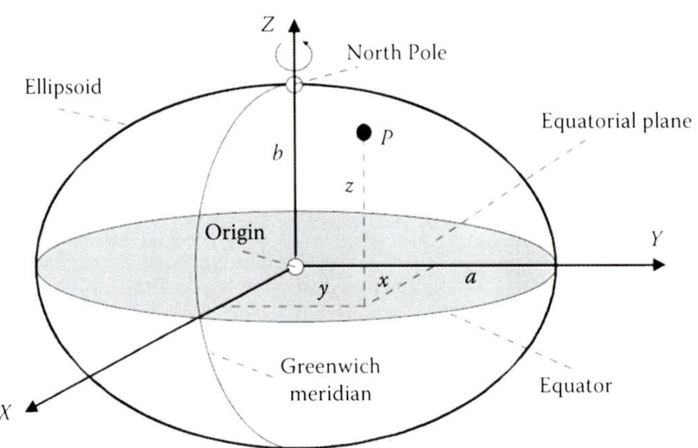

FIGURE 6.13 Illustration of the Cartesian coordinates.

TABLE 6.1
The WGS-84 Reference Ellipsoid

Semimajor axis a (m)	Semiminor axis b (m)	Flattening
6,378,137.00	6,356,752.31	1/298257223563

longitude, and h to the ellipsoidal height, that is, the length of the vertical P line to the ellipsoid.

6.2.2.4 Transformation from Local to Worldwide Reference Ellipsoid

6.2.2.4.1 Geodetic Datum

As a rule, reference systems are generally local rather than geocentric ellipsoids. The relationship between a local (e.g., CH-1903) and a global, geocentric system (e.g., WGS-84) is referred to as the geodetic datum. In the event that the axes of the local and global ellipsoids are parallel, or can be regarded as being parallel for applications within a local area, all that is required for datum transition are three shift parameters, known as the datum shift constants ΔX, ΔY, ΔZ.

A further three angles of rotation φx, φy, φz and a scaling factor m (see Figure 6.15) may have to be added so that the complete transformation formula contains seven parameters. The geodetic datum specifies the location of a local three-dimensional Cartesian coordinate system with regard to the global system.

6.2.2.4.2 Datum Conversion

Converting a datum means by definition converting one three-dimensional Cartesian coordinate system (e.g., WGS-84) into another (e.g., CH-1903) by means of three-dimensional shift, rotation, and extension. In order to effect the conversion, the

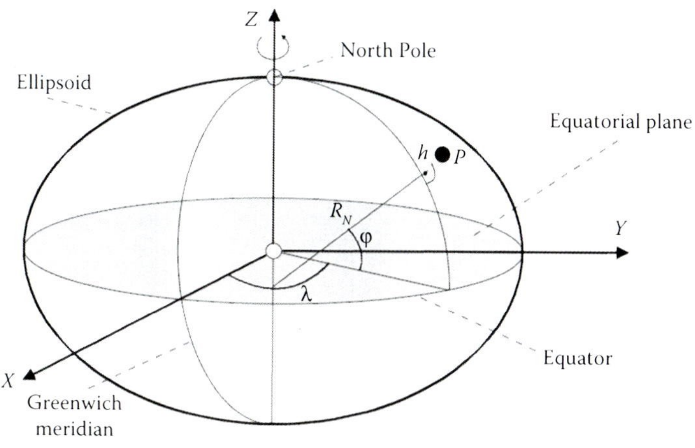

FIGURE 6.14 Illustration of the ellipsoidal coordinates.

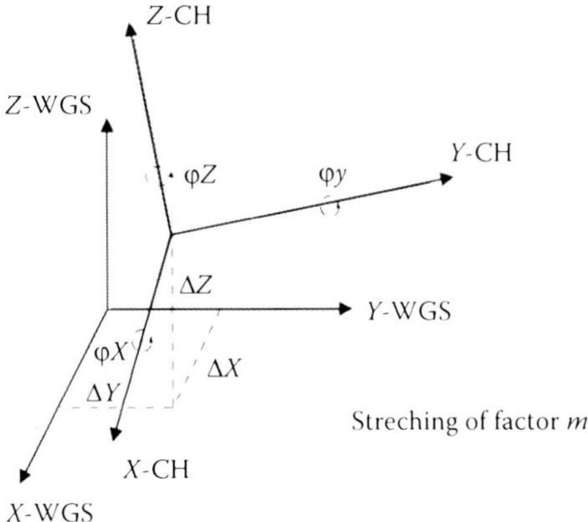

FIGURE 6.15 Geodetic datum.

geodetic datum must be known. Comprehensive conversion formulas can be found in special literature, or conversion can be carried out directly via the Internet. Once conversion has taken place, Cartesian coordinates can be transformed into ellipsoidal coordinates.

6.2.2.5 Converting Coordinate Systems

6.2.2.5.1 Converting Cartesian to Ellipsoidal Coordinates

Cartesian and ellipsoidal coordinates can be converted from one representation to the other. As ϕ and h show up on the right side of the following equations, these equations have to be evaluated iteratively for an accurate solution:

$$e^2 = \frac{a^2 - b^2}{a^2} \tag{6.4}$$

$$R_N = \frac{a}{\sqrt{1 - e^2 \sin^2 \varphi a^2}} \tag{6.5}$$

$$\varphi = \arctan\left[\frac{z}{\sqrt{x^2 + y^2}} * \frac{e^2}{1 - (R_N / (R_N + h))}\right] \tag{6.6}$$

$$\lambda = \tan^{-1}\left(\frac{y}{x}\right) \tag{6.7}$$

$$h = \frac{\sqrt{x^2 + y^2}}{\cos(\varphi)} - R_N \tag{6.8}$$

6.2.2.5.2 Converting Ellipsoidal to Cartesian Coordinates

Ellipsoidal coordinates can be converted into Cartesian coordinates.

$$\begin{aligned} x &= [R_N + h]\cos(\varphi)\cos(\lambda) \\ y &= [R_N + h]\cos(\varphi)\sin(\lambda) \\ z &= [R_N[1 - e^2] + h]\sin(\varphi) \end{aligned} \tag{6.9}$$

6.3 TYPES OF GNSS

6.3.1 Introduction to GPS

All GNSS systems function on the same basic principles. In the following sections, we will explore the different segments of GNSS technology by specifically looking at GPS. GPS is the pioneer and forerunner of GNSS technology and is the most functional GNSS in operation. GPS and GNSS are often used interchangeably, although GPS specifically refers to NAVSTAR GPS, developed by the U.S. Department of Defense and managed by the U.S. Air Force 50th Space Wing. GPS has been fully operational since 1993.

6.3.2 Description of GPS

GPS is comprised of three functional segments (see Figure 6.16):

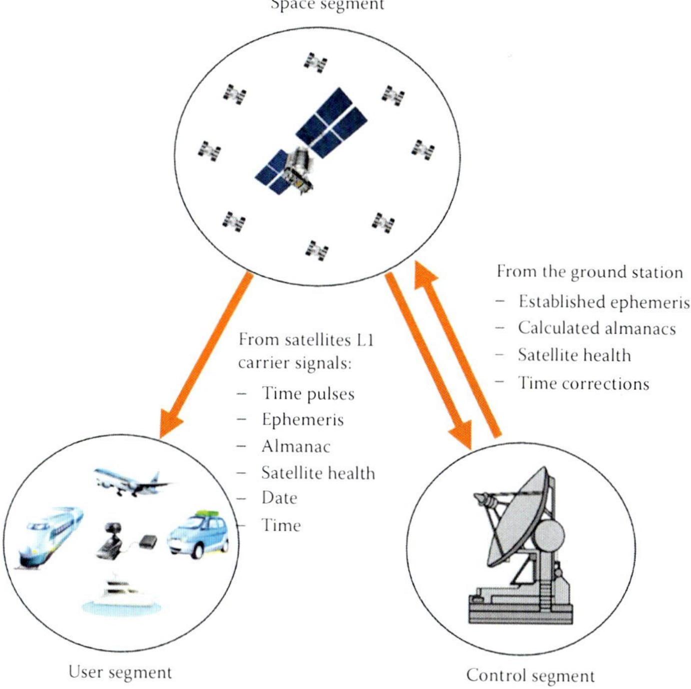

FIGURE 6.16 The three GPS segments.

- Space segment (all operating satellites)
- Control segment (all ground stations involved in the monitoring of the system: master control stations, monitor stations, and ground control stations)
- User segment (all civilian and military users)

As can be seen in Figure 6.16 there is unidirectional communication between the space segment and the user segment. The ground control stations have bidirectional communication with the satellites.

6.3.3 GPS Space Segment

6.3.3.1 Satellite Distribution and Movement

The space segment of GPS consists of up to 32 operational satellites (see Figure 6.17) orbiting the Earth on six different orbital planes (four to five satellites per plane). They orbit at a height of 20,180 km above the Earth's surface and are inclined at 55° to the equator. Each satellite completes its orbit in about 12 hours. Due to Earth's rotation, a satellite will be at its initial starting position above the Earth's surface after approximately 24 hours (23 hours 56 minutes to be precise).

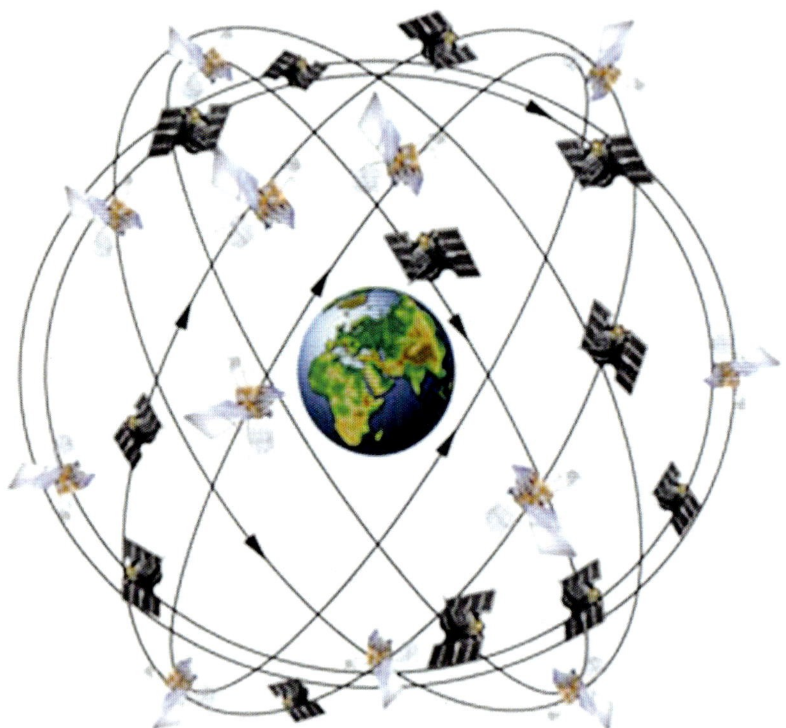

FIGURE 6.17 GPS satellites orbit the Earth on six orbital planes.

6.3.3.2 Satellite Signals in GPS

The following information (the navigation message) is transmitted by the satellite at a rate of 50 bits per second:

- Satellite time and synchronization signals
- Precise orbital data (ephemeris)
- Time correction information to determine the exact satellite time
- Approximate orbital data for all satellites (almanac)
- Correction signals to calculate signal transit time
- Data on the ionosphere
- Information on the operating status (health) of the satellite

The time required to transmit all this information is 12.5 minutes. By using the navigation message, the receiver is able to determine the transmission time of each satellite signal and the exact position of the satellite at the time of transmission.

Each GPS satellite transmits a unique signature assigned to it. This signature consists of a pseudorandom noise (PRN) code of 1023 zeros and ones, broadcast with a duration of 1 ms and continually repeated.

The signature code serves the following two purposes for the receiver:

- Identification (the unique signature pattern identifies the satellite from which the signal originated)
- Signal travel time measurement

6.3.4 GPS Control Segment

The GPS control segment (Operational Control System, OCS) consists of a Master Control Station located in the state of Colorado, five monitor stations (each equipped with atomic clocks and distributed around the globe in the vicinity of the equator), and three ground control stations transmitting information to the satellites.

The most important tasks of the control segment are

- Observing the movement of the satellites and computing orbital data (ephemeris)
- Monitoring the satellite clocks and predicting their behavior
- Synchronizing onboard satellite time
- Relaying precise orbital data received from satellites
- Relaying the approximate orbital data of all satellites (almanac)
- Relaying further information, including satellite condition and clock errors

6.3.5 GPS User Segment

The radio signals transmitted by the GPS satellites take approximately 67 milliseconds to reach a receiver on Earth. As the signals travel at a constant speed (the speed of light c), their travel time determines the exact distance between the satellites and the user.

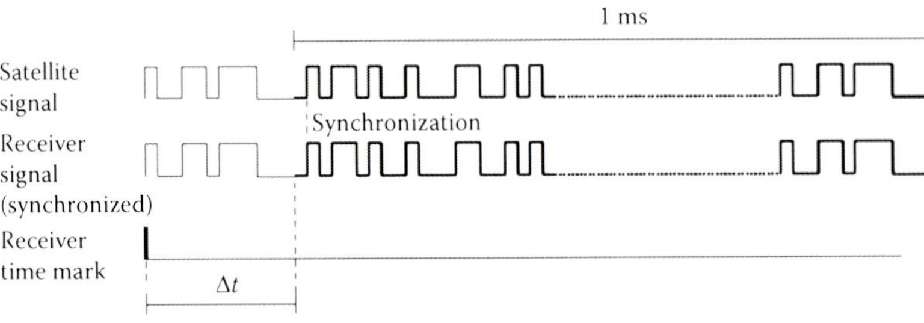

FIGURE 6.18 Measuring signal travel time.

Four different signals are generated in the receiver, each having the same structure as the signals received from the four satellites. By synchronizing the signals generated in the receiver with those from the satellites, the signal time shifts Δt of the four satellites are measured as a time mark (see Figure 6.18). The measured time shifts Δt of all four satellite signals are then used to determine the exact signal travel time. These time shifts multiplied by the speed of light are called pseudoranges.

In order to determine the position of a user, radio communication with four different satellites is required. The distance to the satellites is determined by the travel time of the signals. The receiver then calculates the user's latitude φ, longitude λ, altitude h, and time t from the pseudoranges and known positions of four satellites. Expressed in mathematical terms, this means that the four unknown variables φ, λ, h, and t are determined from the distance and known position of these four satellites.

6.3.6 GLONASS: The Russian System

GLONASS is an abbreviation for the GNSS currently operated by the Russian Defense Ministry. The designation GLONASS stands for Global Navigation Satellite System. The program was started by the former Soviet Union, and today is under the jurisdiction of the Commonwealth of Independent States (CIS). The first three test-satellites were launched into orbit on October 12, 1982.

The most important facts of the GLONASS system are

- GLONASS constellation status on February 18, 2017: total satellites in constellation: 27 SC; operational: 24 SC; under check by the Satellite Prime Contractor: 1 SC; Spares: 1 SC; in flight tests phase: 1 SC.
- Three orbital levels with an angle of 64.8° from the equator (this is the highest angle of all GNSS and allows better reception in polar regions).
- Orbital altitude of 19,100 km and orbital period of 11 h 15.8 min.
- Every GLONASS satellite transmits two codes (C/A and P-Code) on two frequencies. Every satellite transmits the same code but at different frequencies in the vicinity of 1602 MHz (L1 Band) and 1246 MHz (L2 Band). The frequencies can be determined through the following formula (k is the frequency channel of the satellite under consideration):

- Frequency in L1 Band: f1 = 1602 MHz + $k \times (9/16)$ MHz
- Frequency in L2 Band: f2 = 1246 MHz + $k \times (7/16)$ MHz

6.3.7 Overview of Galileo

Galileo is the European GNSS being developed by the European Union (EU), in close cooperation with the European Space Agency (ESA). Galileo will consist of a constellation of 30 satellites on three circular orbits at an altitude of 23,222 km above the Earth. These satellites are to be supported by a worldwide network of ground stations.

The key arguments from the perspective of the EU for introducing Galileo are

- To attain independence from the United States.
- To have a precise navigation system. The open service (OS) is expected to provide a precision of approximately 4–15 m. Critical security services should have a precision of 4–6 m. Sensitivity to multipath reception will also be reduced. This improvement will be achieved through the application of BOC (Binary Offset Carrier) and MBOC (Multiplexed Binary Offset Carrier) modulation. GPS will also introduce BOC and MBOC when it is modernized.
- To have a purely civilian navigation system. Galileo is being conceived and implemented according to civilian criteria. For some services, Galileo will offer a guarantee of function.
- Providing more services. Galileo will offer five different functions.
- Offer a search and rescue function. Search and rescue (SAR) functions are already being offered by other organizations. New with Galileo is that an alarm can be acknowledged.
- Increased security through integrity messages. Galileo will be more reliable in that it includes an integrity message. This will immediately inform users of errors that develop. On top of this is a guarantee of availability. For the open service, there will be neither the availability guarantee nor the integrity messages. These services are only available through the European Geostationary Navigation Overlay Service (EGNOS).
- Creation of employment.
- Attain GNSS know-how. With Galileo, Europe wants to acquire expertise and provide the domestic industry with a sustainable growth in competence. For example, the atomic clocks used by Galileo are to be manufactured in Europe.
- To improve the worldwide coverage of satellite signals. Galileo will offer better reception than GPS to cities located in higher latitudes. This is possible because the Galileo satellites have orbits at an angle of 56° from the equator as well as an altitude of 23,616 km. In addition, modern GNSS receivers are able to evaluate GPS and Galileo signals. This multiplies the number of visible satellites from which signals can be received, increasing the level of coverage and the accuracy.

6.3.8 The Chinese System: BeiDou 1 and BeiDou 2/Compass

6.3.8.1 Current System: BeiDou 1

Between 2000 and 2007, China placed four geostationary satellites into service for the local BeiDou system. The satellites (BeiDou-1A to BeiDou-1D) transmit over China. Position is interactively determined between satellites and navigation receivers. The signals are transmitted and received in an iterative method:

1. A signal is transmitted from the navigation receiver to four GEO (geostationary Earth orbit) satellites.
2. All four satellites receive the signal.
3. All four satellites transmit the exact time of signal reception to a ground station.
4. The ground station calculates the longitude and latitude of the navigation receiver and determines its elevation.
5. The ground station transmits the position to the GEO satellites.
6. The GEO satellites transmit the position to the navigation receiver.

6.3.8.2 Future System: BeiDou 2/Compass

China is developing a GNSS currently known as BeiDou 2 or Compass, CNSS (Compass Navigation Satellite System). The system will consist of 5 GEO and 30 MEO (medium Earth orbit) satellites. The GEO satellites will expand the entire system similar to SBAS. BeiDou/Compass should offer two navigation services:

- Open service, with position accuracy of 10 m, velocity accuracy of 0.2 m/s, and time precision of 50 ns.
- Service for authorized users. This service should be more reliable than the open service.

The MEO satellites will be distributed over six orbits. The first of these satellites was launched into orbit in April 2007. The time of completion of the total system is around 2020.

6.4 DIFFERENTIAL GPS AND SBAS (SATELLITE-BASED AUGMENTATION SYSTEMS)

Although originally intended for military purposes, today GPS is used primarily for civil applications such as surveying, navigation, positioning, measuring velocity, determining time, and monitoring. GPS was not initially conceived for applications demanding high precision, security measures, or for indoor use.

For the use of GPS for civil applications:

- To increase the accuracy of positioning, differential GPS (DGPS) was introduced.
- To improve the accuracy of positioning and the integrity (reliability, important for security applications), SBAS (Satellite-Based Augmentation

Systems) such as EGNOS and WAAS (Wide Area Augmentation System) were implemented.

6.4.1 Sources of GPS Error

The positioning accuracy of approximately 10 m is not sufficient for all applications. In order to achieve an accuracy of 1 m or better, additional measures are necessary. Different sources can contribute to the total error in GPS measurements. These values should be viewed as typical averages and can vary from receiver to receiver.

Sources for errors include:

- Ephemeris data—The satellite position at the time of the signal emission is, as a general rule, only known to be accurate up to approximately 1 to 5 m.
- Satellite clocks—Although each satellite includes four atomic clocks, the time base contains offsets. A time error of 10 ns is reached at an oscillator stability of approximately 10^{-13} per day. A time error of 10 ns immediately results in a distance error of about 3 m.
- Effect of the ionosphere—The ionosphere is an atmospheric layer situated between 60 and 1000 km above the Earth's surface. The gas molecules in the ionosphere are heavily ionized. The ionization is mainly caused by solar radiation (only during the day!). Signals from the satellites travel through a vacuum at the speed of light. In the ionosphere, the velocity of these signals slows down and therefore can no longer be viewed as constant. The level of ionization varies depending on time and location, and is strongest during the day and at the equator. If the ionization strength is known, this effect can, to a certain extent, be compensated with geophysical correction models. Furthermore, given that the change in the signal velocity is frequency dependent, this can additionally be corrected by the use of dual-frequency GPS receivers.
- Effect of the troposphere—The troposphere is the atmospheric layer located between 0 and 15 km above the Earth's surface. The cause of the error here is the varying density of the gas molecules and the air humidity. The density decreases as the height increases. The increase in density or humidity slows the speed of the satellite signals. In order to correct this effect, a simple model is used based on the standard atmosphere (P) and temperature (T):

  H = height (m)

  $T = 288.15\text{ K} - 6.5 \times 10^{-3} \times h\text{ (K)}$

  $P = 1013.25\text{ mbar } (T/288.15\text{ K})^{5.256}\text{ (mbar)}$
- Multipath—GPS signals can be reflected from buildings, trees, mountains, and so forth, then make a detour before arriving at the receiver. The signal is distorted due to interference. The effect of multipath can be partially compensated by the selection of the measuring location (free of reflections), a good antenna, and the measuring time (Figure 6.19).
- Effect of the receiver—Further errors are produced due to GPS receiver measurement noise and time delays in the receiver. Advanced technologies can be used to reduce this effect.
- Effect of the satellite constellation, including shadowing.

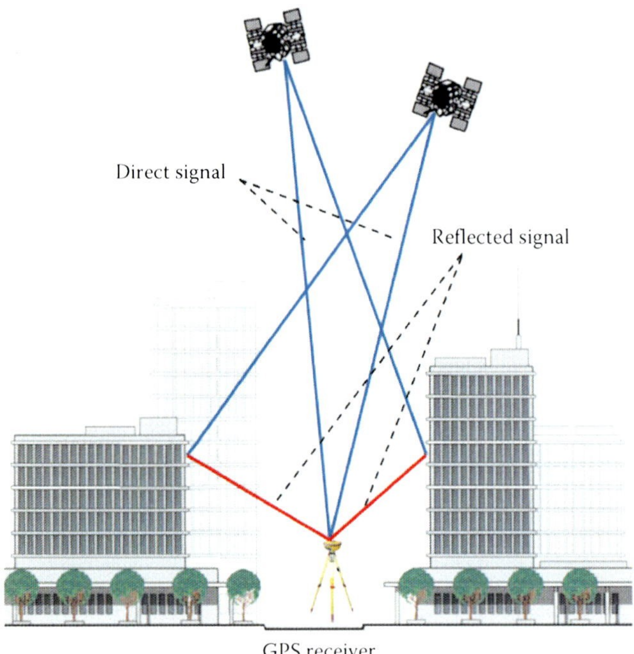

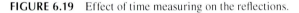

FIGURE 6.19 Effect of time measuring on the reflections.

6.4.2 Possibilities for Reducing the Measurement Error

Reducing the effect of measurement errors can considerably increase the positioning accuracy. Different approaches are used for reducing the measurement error and are often combined. The process most frequently used is compensation of ionospheric influences through dual-frequency measurement. The ionosphere has the greatest influence on measurement errors. If a radio signal is transmitted through the ionosphere, it is slowed more heavily at lower frequencies. By using two different signal frequencies, for example, L1/L2, the effect of the ionosphere can largely be compensated for. Since Galileo and GPS (following modernization) will transmit the civil signal on at least two frequencies, the principle of compensation will be more closely adhered to.

Signal transmission times increase depending on the strength of ionization. The ionospheric effect on transmission time increases with lower frequencies. The influence on the transmission time occurs as a square of the frequency.

If the signal is slower, then a longer distance between the satellite and the receiver is assumed. The measurement error of this pseudorange (PSR) and its dependence on frequency and ionization strength is provided in Figure 6.20.

Since every satellite signal is transmitted through a different area of ionization, the PSR measurement error is different for every satellite. It is, therefore, important to compensate for these errors. If a satellite transmits navigation information on two frequencies (f_1 and f_2), it is possible to determine the PSR measurement error (ΔPSR_1) for frequency f_1 by using the following formula:

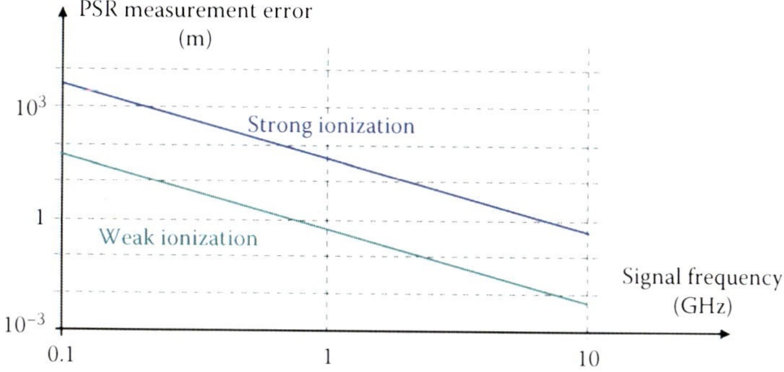

FIGURE 6.20 PSR measurement error and its dependence on ionization and frequency.

$$\Delta PSR_1 = \left(\frac{(f_2)^2}{(f_2)^2 - (f_1)^2} \right) \cdot (PSR_1 - PSR_2) \tag{6.10}$$

PSR_1 and PSR_2 are the measured pseudoranges for frequencies f_1 and f_2. The calculated measurement error (ΔPSR_i) can be used for the correction of the PSR_i value in the equation of single point positioning:

$$\begin{bmatrix} \Delta x \\ \Delta y \\ \Delta z \\ \Delta t_0 \end{bmatrix} = \begin{bmatrix} \frac{X_{Total} - X_{Sat\_1}}{R_{Total\_1}} & \frac{Y_{Total} - Y_{Sat\_1}}{R_{Total\_1}} & \frac{Z_{Total} - Z_{Sat\_1}}{R_{Total\_1}} & c \\ \frac{X_{Total} - X_{Sat\_2}}{R_{Total\_2}} & \frac{Y_{Total} - Y_{Sat\_2}}{R_{Total\_2}} & \frac{Z_{Total} - Z_{Sat\_2}}{R_{Total\_2}} & c \\ \frac{X_{Total} - X_{Sat\_3}}{R_{Total\_3}} & \frac{Y_{Total} - Y_{Sat\_3}}{R_{Total\_3}} & \frac{Z_{Total} - Z_{Sat\_3}}{R_{Total\_3}} & c \\ \frac{X_{Total} - X_{Sat\_4}}{R_{Total\_4}} & \frac{Y_{Total} - Y_{Sat\_4}}{R_{Total\_4}} & \frac{Z_{Total} - Z_{Sat\_4}}{R_{Total\_4}} & c \end{bmatrix}^{-1} \times \begin{bmatrix} PSR_1 - \Delta PSR_1 - R_{Total\_1} \\ PSR_2 - \Delta PSR_2 - R_{Total\_2} \\ PSR_3 - \Delta PSR_3 - R_{Total\_3} \\ PSR_4 - \Delta PSR_4 - R_{Total\_4} \end{bmatrix} \tag{6.11}$$

Geophysical correction models are another compensation method that is primarily used for the compensation of the effect of the ionosphere and troposphere. Correction factors are only useful if applied to a specified and limited area.

Differential GPS (DGPS) provides improved accuracy of positioning. By comparing with one or several base stations, various errors can be corrected. The evaluation of the correction data available from these stations can take place either during postprocessing or in real time (RT). Real-time solutions (RT DGPS) require data communication between the base station and the mobile receiver. DGPS employs a variety of different processes:

- RT DGPS, normally based on the RTCM SC104 standard
    - DGPS derived from signal travel time delay measurement (pseudorange corrections, achievable accuracy approximately 1 m)
    - DGPS derived from the phase measurement of the carrier signal (achievable accuracy approximately 1 cm)
- Postprocessing (subsequent correction and processing of the data)

Choice of location and of the measurement time for improving the "visibility" or line of sight contact to the satellites also provide possibilities for reducing the measurement error.

6.4.3 Differential GPS (DGPS) Based on Signal Travel Time Delay Measurement

The principle of DGPS based on signal travel time measurement (pseudorange or C/A code measurement) is very simple. A GPS reference station is located at a known and accurately surveyed point. The GPS reference station determines its GPS position using four or more satellites. Given that the position of the GPS reference station is precisely known, the deviation of the measured position to the actual position and, more important, the measured pseudorange to each of the individual satellites can be calculated. These variations are valid for all the GPS receivers around the GPS reference station in a range of up to 200 km. The satellite pseudoranges can thereby be used for the correction of the measured positions of other GPS receivers (Figure 6.21). The differences are either transmitted immediately by radio or used afterward for correction after carrying out the measurements.

It is important that the correction should be based on the satellite pseudorange values and not on the specific deviation in position of the GPS reference station. Deviations are based on the pseudoranges to the specific satellites, and these vary depending on position as well as which satellites are used. A correction based simply on the positional deviation of the reference base station fails to take this into account and will lead to false results.

6.4.3.1 Detailed Description of How It Runs

The error compensation is carried out in three phases:

1. Determination of the correction values at the reference station
2. Transmission of the correction values from the reference station to the GPS user
3. Compensation for the determined pseudoranges to correct the calculated position of the GPS user

6.4.3.2 Definition of the Correction Factors

A reference station with exactly known position measures the L1 signal travel time to all visible GPS satellites (Figure 6.22) and uses these values to calculate

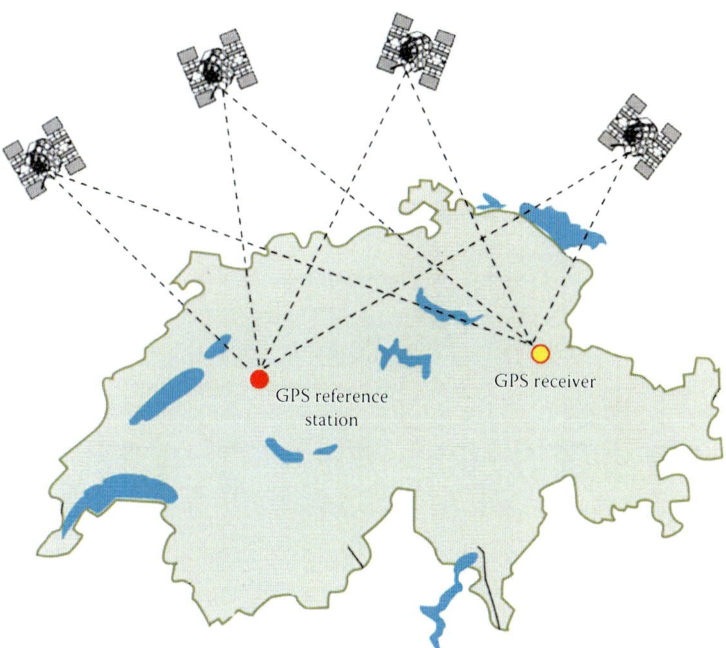

FIGURE 6.21 Principle of DGPS with a GPS base station.

its position relative to the satellites. These measured values will typically include errors. Since the real position of the reference station is known, the actual distance (nominal value) to each GPS satellite can be calculated. The difference between the nominal and the measured distances can be calculated by a simple subtraction and corresponds to a correction factor. These correction factors are different for all GPS

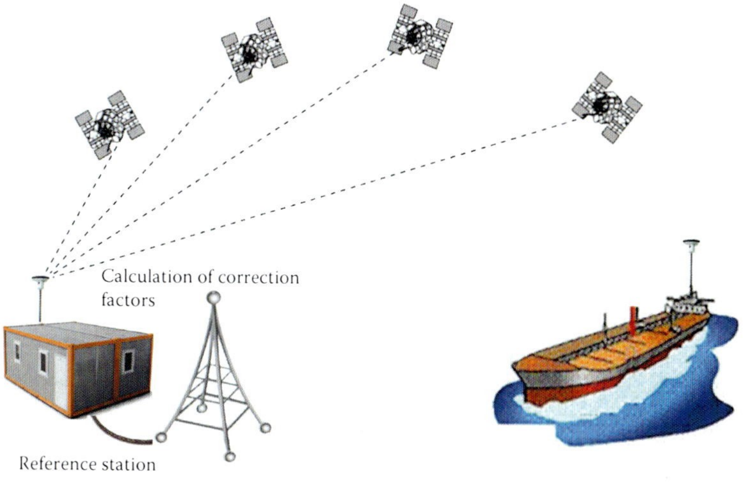

FIGURE 6.22 Determination of the correction factors.

satellites and are also applicable to GPS users within a radius of several hundred kilometers.

6.4.3.3 Transmission of the Correction Values

Given that the correction values can be used by other GPS users within a large area to compensate for the measured pseudoranges, they are immediately transmitted by using a suitable medium (telephone, radio, etc.) (Figure 6.23).

6.4.3.4 Correction of the Measured Pseudoranges

After receiving the correction values, the GPS user can compensate for the pseudoranges in order to determine the actual distance to the satellites (Figure 6.24). These actual distances can then be used to calculate the exact position of the user. All errors, which are not caused by receiver noise and multipath reception, can be cleared this way.

6.4.3.5 DGPS Based on Carrier Phase Measurement

The DGPS accuracy of 1 m achieved by measuring signal travel time is not enough for some requirements such as solving survey problems. In order to obtain a precision within millimeters, the carrier phase of the satellite signal must be evaluated.

The wavelength λ of the carrier wave is approximately 19 cm. The distance to a satellite can be determined as shown in Figure 6.25. Since N is unknown, the phase measurement is ambiguous. By observing several satellites at different times and continually comparing results from users and reference station receivers (during or after the measurement), the position can be calculated using an extensive series of mathematical equations to an accuracy of a few millimeters.

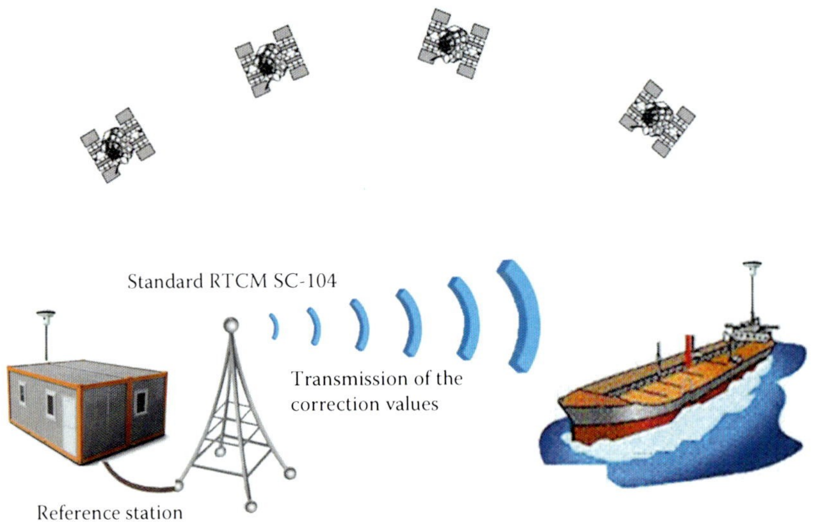

FIGURE 6.23 Transmission of the correction factors.

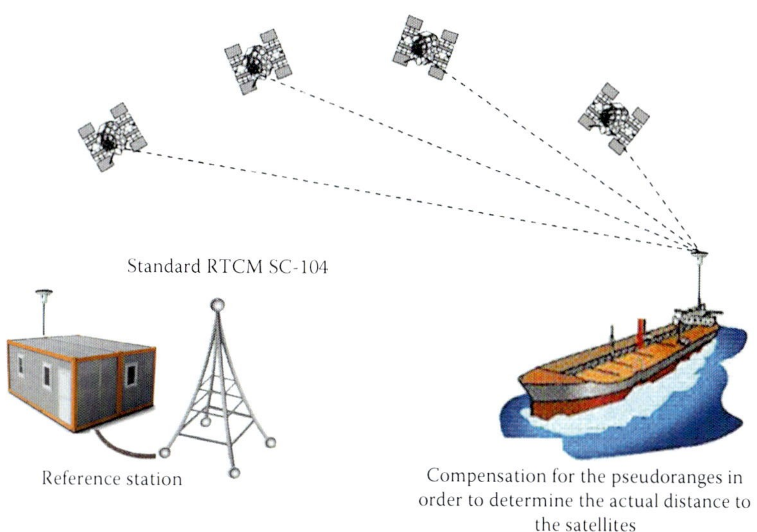

FIGURE 6.24 Correction of the measured pseudoranges.

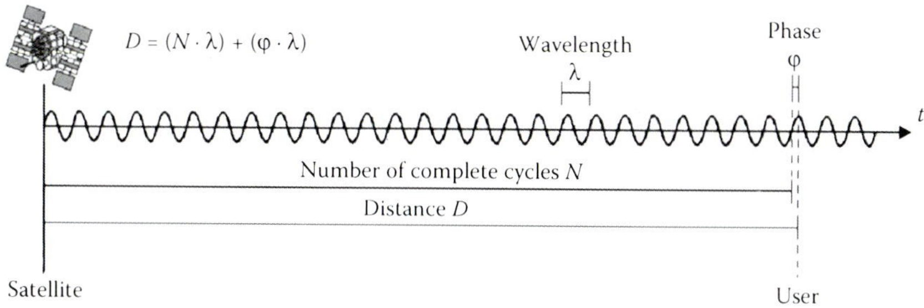

FIGURE 6.25 Principle of the phase measurement.

6.4.3.6 DGPS Postprocessing (Signal Travel Time and Phase Measurement)

DGPS postprocessing implements the determined correction factors by using appropriate software after carrying out field measurements. Reference data is either obtained from private reference stations or from publicly accessible server systems. The disadvantage of this is that problems with the field data (e.g., poor satellite reception and damaged files) are sometimes not detected until after the correction factors are calculated and broadcasted, requiring repetition of the whole process.

6.4.3.7 DGPS Classification according to the Broadcast Range

Various DGPS services available are categorized according to the broadcast range of the correction signals:

- Local DGPS—Local Area Augmentation System (LAAS). These are sometimes called Ground-Based Augmentation Systems (GBAS).
- Regional DGPS.
- Wide Area DGPS (WADGPS) or Satellite-Based Augmentation Systems (SBAS)—Employs satellites to transmit DGPS correction data. In these cases, not only are single reference stations, but whole networks of reference stations are used.

6.4.3.8 Standards for the Transmission of Correction Signals

DGPS broadcasters transmit the signal travel time and carrier phase corrections. For most GBAS and some satellite-based WADGPS systems (LandStar-DGPS, MSAT, Omnistar, or Starfire), the DGPS correction data is transmitted according to the Radio Technical Commission for Maritime Services Special Committee 104 (RTCM SC-104) standard. Typically, the receiver must be equipped with a service-specific decoder in order to receive and process the data.

Satellite-Based Augmentation Systems such as WAAS, EGNOS, and MSAS (Multifunctional Satellite Augmentation System) use the RTCA DO-229 standard. Since RTCA (Radio Technical Commission for Aeronautics) frequencies and data formats are compatible with those of GPS, modern GNSS receivers can calculate RTCA data without the use of additional hardware, in contrast to RTCM (Figure 6.26).

Table 6.2 lists the standards used for DGPS correction signals as well as the references pertaining specifically to GNSS.

6.4.3.9 Overview of the Different Correction Services

Figure 6.27 presents classifications of different GNSS correction services.

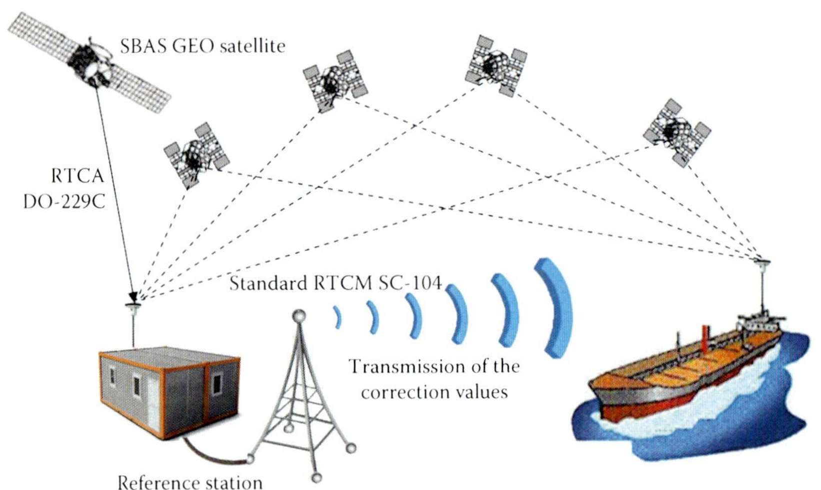

FIGURE 6.26 Comparison of DGPS systems based on RTCM and RTCA standards.

TABLE 6.2
Standards for DGPS Correction Signals

Standard	References Pertaining to GNSS
RTCM SC 104: Radio Technical Commission for Maritime Services, Special Committee 104	RTCM Recommended Standards for Differential Navstar GPS Service, Version 2.0 and 2.1 Recommended Standards for Differential GNSS Service, Version 2.2 and 2.3
RTCA: Radio Technical Commission for Aeronautics	DO-229C, Minimum Operational Performance Standards for Global Positioning System/Wide Area Augmentation System Airborne Equipment

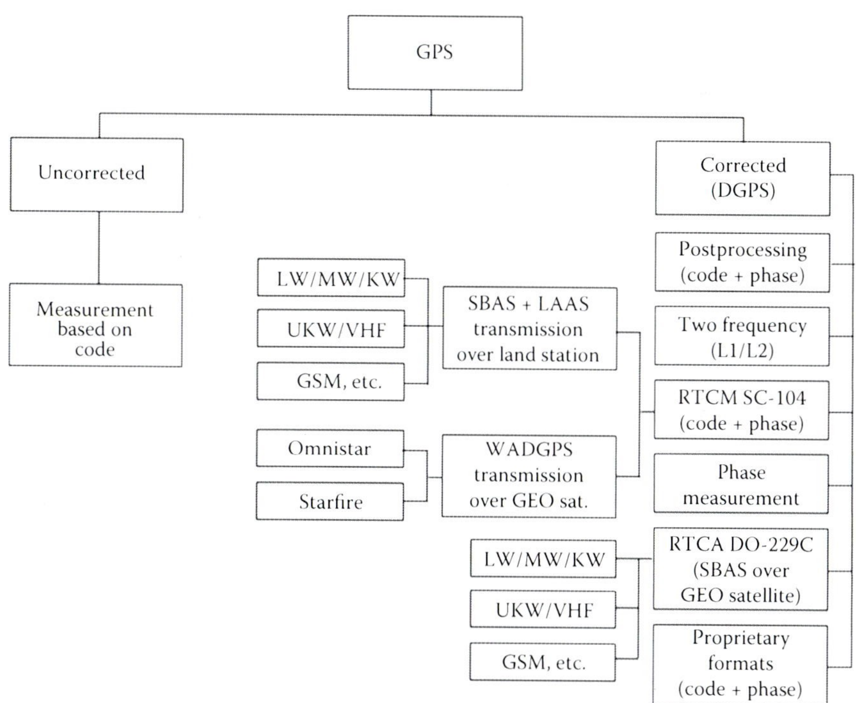

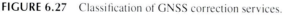

FIGURE 6.27 Classification of GNSS correction services.

6.4.4 SBAS

SBAS are used to enhance GPS, GLONASS, and Galileo (once it is operational) functions. Correction and integrity data for GPS or GLONASS is broadcast from geostationary satellites over the GNSS frequency.

6.4.4.1 Most Important SBAS Functions

SBAS is a considerable improvement compared to GPS because the positioning accuracy and the reliability of the positioning information are improved. SBAS, in contrast to GPS, delivers additional signals broadcast from different geostationary satellites. SBAS advantages include:

- Increased positioning accuracy using correction data—SBAS provides differential correction data with which the GNSS positioning accuracy is improved. First, the ionospheric error, which arises due to the signal delays in the ionosphere, has to be corrected. The ionospheric error varies with the time of day and is different from region to region. To ensure that the data is continentally valid, it is necessary to operate a complicated network of earth stations to be able to calculate the ionospheric error. In addition to the ionospheric values, SBAS passes on the correction information concerning the satellite position location (ephemeris) and time measurement.
- Increased integrity and security—SBAS monitors each GNSS satellite and notifies the user of a satellite error or breakdown within a short warning latency of 6 s. This yes/no information is transmitted only if the quality of the received signals remains below specific limits.
- Increased availability through the broadcasting of navigation information—SBAS geostationary satellites emit signals, which are similar to the GNSS signals although missing the accurate time data. A GNSS receiver can interpret position from these signals using a procedure known as "pseudoranging."

6.4.4.2 Overview of Existing and Planned SBAS

Although all SBAS include very large regions (e.g., Europe), it must be ensured that they are compatible with one another (interoperability) and that the SBAS providers cooperate with one another and agree on their approach. Compatibility is guaranteed by using the RTCA DO-229C standard. Currently, compatible SBAS are operated or being developed for the areas identified in Figure 6.28.

- North America (WAAS, Wide Area Augmentation System)—The U.S. Federal Aviation Administration (FAA) is developing WAAS, which covers the United States, Canada, and Mexico (see Figure 6.29). WAAS operates over two satellites (Anik F1R und Galaxy 15) located at 133°W and 107°W.

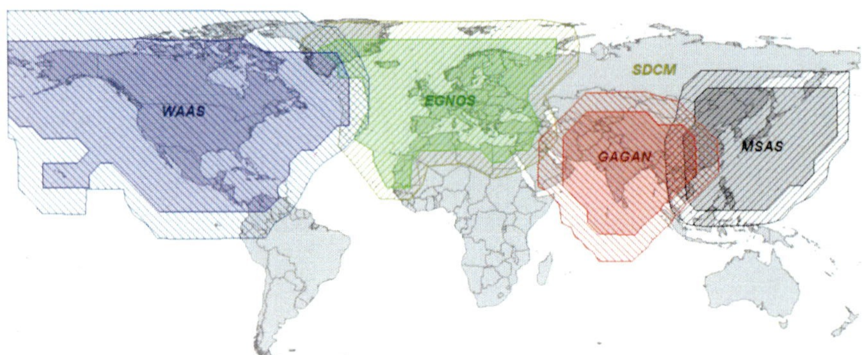

FIGURE 6.28 Position and coverage of WAAS, EGNOS, GAGAN, and MSAS.

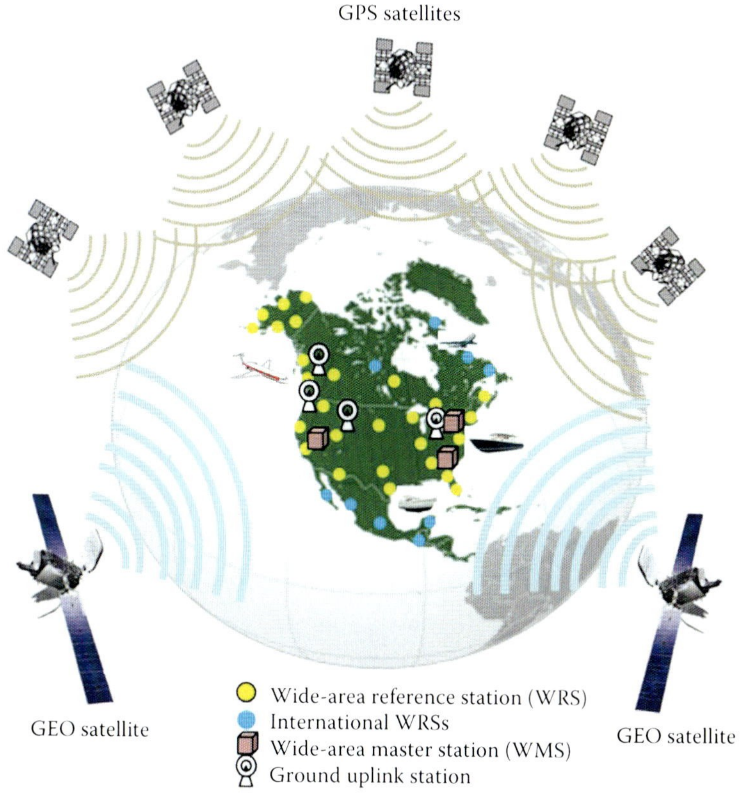

FIGURE 6.29 WAAS area of coverage.

- Europe (EGNOS, European Geostationary Navigation Overlay Service)—EGNOS was developed by the European Space Agency, the European Commission and EUROCONTROL. It supplements the GPS, GLONASS, and Galileo satellite navigation systems by reporting on the reliability and accuracy of their positioning data and sending out corrections. The official start of operations was announced by the European Commission on October 1, 2009. The system was certified for use in safety of life applications in March 2011. An EGNOS Data Access Service became available in July 2012.
- Japan (QZSS, Quasi-Zenith Satellite System)—The Japan regional satellite system. QZSS is three-satellite regional time transfer system and the satellite-based augmentation system for the Global Positioning System that would be receivable within Japan. The first satellite "Michibiki" was launched on September 11, 2010. QZSS was commissioned in 2013. In March 2013, Japan's Cabinet Office announced the expansion of the Quasi-Zenith Satellite System from three satellites to four. The $526 million contract with Mitsubishi Electric for the construction of three satellites is slated

for launch before the end of 2017. The basic four-satellite system is planned to be operational in 2018.
- India (GAGAN, GPS, and GEO Augmented Navigation)—The Indian Space Research Organization (ISRO) is developing a system, which will be compatible with the other SBAS systems. In 2008, GAGAN test signals were transmitted by Inmarsat 4F1 IOR. Satellites GSAT-8 and GSAT-10 began to operate as part of GAGAN in 2011 and 2012, respectively. The new GAGAN satellite GSAT-15 (85.0° E) has started operating from 2015.
- Russia (SDCM, System for Differential Correction and Monitoring)—Russia is developing a system for its territory to control GPS and GLONASS signals using various monitoring stations. GEO satellites will transmit correction and integrity signals for GPS and GLONASS over Russian territory.

6.5 GNSS RECEIVERS

6.5.1 Basics of GNSS Handheld Receivers

A GNSS receiver can be divided into the following main stages (Figure 6.30):

- Antenna—The antenna receives extremely weak satellite signals on a frequency of 1572.42 MHz. Signal output is around −163 dBW. Some (passive) antennas have a 3 dB gain.
- LNA 1—This low-noise amplifier (LNA) amplifies the signal by approximately 15–20 dB.
- RF filter—The GNSS signal bandwidth is approximately 2 MHz. The RF filter reduces the affects of signal interference. The RF stage and signal processor actually represent the special circuits in a GNSS receiver and are adjusted to each other.
- RF stage—The amplified GNSS signal is mixed with the frequency of the local oscillator. The filtered intermediate frequency (IF) signal is maintained at a constant level with respect to its amplitude and digitalized via amplitude gain control (AGC).
- IF filter—The intermediate frequency is filtered out using a bandwidth of several megahertz. The image frequencies arising at the mixing stage are reduced to a permissible level.
- Signal processor—Up to 16 different satellite signals can be correlated and decoded at the same time. Correlation takes place by constant comparison with the C/A code. The RF stage and signal processor are simultaneously switched to synchronize with the signal. The signal processor has its own time base (real-time clock, RTC). All the data ascertained is broadcast (particularly signal transit time to the relevant satellites determined by the correlator) and this is referred to as the source data. The signal processor can be programmed by the controller via the control line to function in various operating modes.

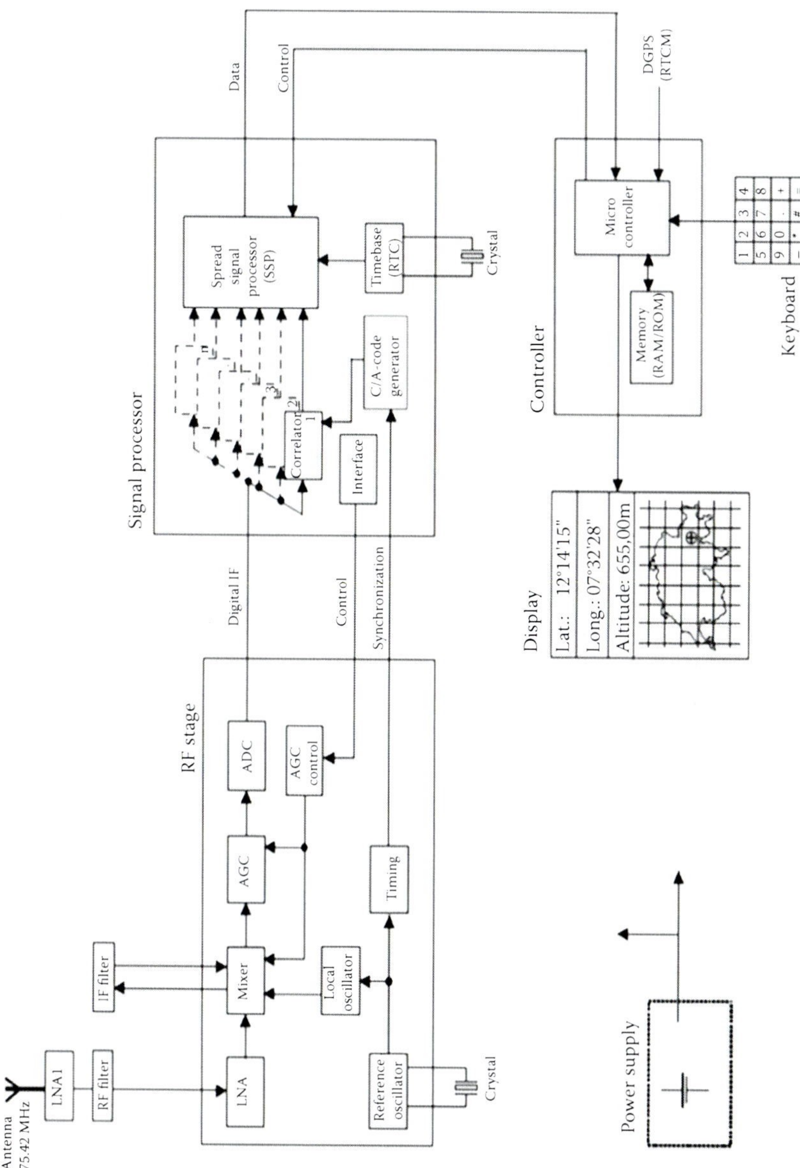

FIGURE 6.30 Simplified block diagram of a GNSS receiver.

- Controller—Using the source data, the controller calculates position, time, speed, course, and so on. It controls the signal processor and relays the calculated values to the display. Important information (such as ephemeris and the most recent position) is decoded and saved in RAM. The program and the calculation algorithms are saved in ROM.
- Keyboard—Using the keyboard, the user can select which coordinate system he wishes to use and which parameters (e.g., number of visible satellites) should be displayed.
- Display—The position calculated (longitude, latitude, and height) must be made available to the user. This can either be displayed using a seven-segment display or shown on a screen using a projected map. The positions determined can be saved and whole routes can be recorded.
- Power supply—The power supply delivers the necessary operational voltage to all electronic components.

Table 6.3 shows a comparison in technical specifications of contemporary GNSS handheld receivers used for geodetic survey. Please keep in mind that their performance in positioning accuracy is provided when differential correction is applied.

6.5.2 Basic Design of a GNSS Module

GNSS modules have to evaluate weak antenna signals from at least four satellites in order to determine a correct three-dimensional position. A time signal is also often emitted in addition to longitude, latitude, and height. This time signal is synchronized with UTC (Coordinated Universal Time). From the position determined and knowing the exact time, additional physical variables such as speed and acceleration can also be calculated. The GNSS module issues information on the constellation, satellite condition, the number of visible satellites, and so on.

Figure 6.31 shows a typical block diagram of a GNSS module. The signals received (1575.42 MHz) are preamplified and transformed to a lower intermediate frequency. The reference oscillator provides the necessary carrier wave for frequency conversion, along with the necessary clock frequency for the processor and correlator. The analog intermediate frequency is converted into a digital signal by means of an analog-to-digital converter.

Signal travel time from the satellites to the GNSS receiver is determined by correlating PRN pulse sequences. The satellite PRN sequence must be used to establish this time, otherwise there is no correlation maximum. Data is recovered by mixing it with the correct PRN sequence. At the same time, the useful signal is amplified above the interference level 104. Up to 16 satellite signals are processed simultaneously. A signal processor carries out the control and generation of PRN sequences and the recovery of data. Calculating and saving the position, including the variables derived from this, is carried out by a processor with a memory facility.

Table 6.4 compares state-of-the-art GNSS receivers with embedded antennas used as geodetic equipment. Please bear in mind that their performance in positioning accuracy is computed when differential correction is applied.

TABLE 6.3
Comparison of Handheld Geodetic GNSS Receivers

Specifications	UniStrong MG8	Trimble GeoXR	Topcon GRS-1	Stonex S7
GPS	Yes	Yes (L1 + L2)	Yes (L1 + L2)	Yes
GLONASS	Yes	Yes (L1 + L2)	Yes (L1 + L2)	Yes
Galileo	Yes	No	No	Yes
BeiDou	Yes	No	No	Yes
SBAS	Yes	Yes	Yes	Yes
Number of channels	372	220	72	120
Horizontal accuracy in static surveying	5 mm + 0.5 mm/km	3 mm + 0.5 mm/km	3 mm + 0.5 mm/km	5 mm + 0.5 mm/km
Vertical accuracy in static surveying	10 mm + 0.5 mm/km	3.5 mm + 1 mm/km	5 mm + 0.5 mm/km	10 mm + 0.5 mm/km
Horizontal accuracy in RTK surveying	10 mm + 1 mm/km	10 mm + 1 mm/km	10 mm + 1 mm/km	10 mm + 1 mm/km
Vertical accuracy in RTK surveying	20 mm + 1 mm/km	15 mm + 1 mm/km	15 mm + 1 mm/km	20 mm + 1 mm/km
Internal memory	256 Mb	512 Mb	256 Mb	256 Mb
Water/dustproof	IP66-IP67	IP65	IP 66	IP 65
Measurement update	1 Hz	1 Hz	Up to 10 Hz	1 Hz
Bluetooth	Yes	Version 2.1 + EDR	Yes	Yes
Wi-Fi	802.11 b/g	802.11 b/g	802.11 b/g	802.11 b/g
Operating temperature	From −20 to +50	From −20 to +50	From −20 to +50	From −20 to +50
Weight	0.850 kg	0.925 kg	0.77 kg	0.850 kg
Dimension	99 × 134 × 56 mm	99 × 234 × 56 mm	93 × 215 × 53	99 × 134 × 56 mm

6.6 GNSS APPLICATIONS

Using GNSS, the following two values can be determined anywhere on Earth:

- Exact position (longitude, latitude, and height coordinates) accurate to within a range of 20 m to approximately 1 mm
- Precise time (UTC) accurate to within a range of 60 ns to approximately 1 ns

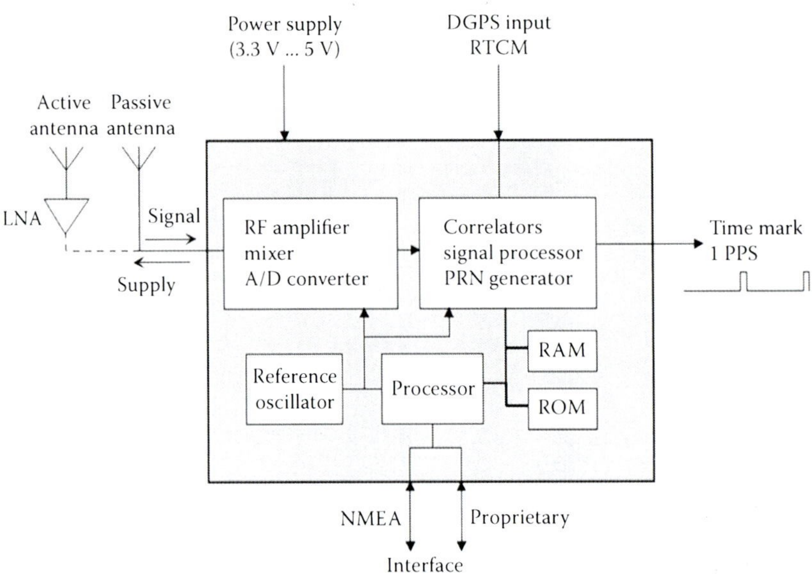

FIGURE 6.31 Typical block diagram of a GNSS module.

In addition, other values can also be determined, such as

- Speed
- Acceleration
- Course
- Local time
- Range measurements

The established fields for GNSS usage are surveying, shipping, and aviation. However, satellite navigation is currently enjoying a surge in demand for location-based services (LBS) and systems for the automobile industry. Applications such as automatic vehicle location (AVL) and the management of vehicle fleets also appear to be on the rise. In addition, GNSS is increasingly being utilized in communications technology. For example, the precise GNSS time signal is used to synchronize telecommunications networks around the world. Since 2001, the U.S. Federal Communications Commission (FCC) has required that when Americans call 911 in an emergency that their position be automatically determined to within approximately 125 m. This law, known as E-911 (Enhanced 911), necessitates that mobile telephones be upgraded with this new technology.

In the leisure industry, GNSS is becoming increasingly widespread and important. For example, whether hiking, hunting, mountain biking, or windsurfing across Lake Constance in southern Germany, a GNSS receiver provides invaluable information for a great variety of situations.

GNSS can essentially be used anywhere on Earth where satellite signal reception is possible and knowledge of position is of benefit.

TABLE 6.4
Comparison of Geodetic GNSS Receivers with Smart Antenna

Specifications	UniStrong G970	Altus APS-3L	Acnovo GX9	South S82-V	Trimble R4-3	Topcon Hiper V	Geomax Zenith 25	Leica Viva GS15
GPS	Yes	Yes	Yes	Yes	Yes	Yes	Yes	Yes
GLONASS	Yes	Yes	Yes	Yes	Yes	Yes	Yes	Yes
Galileo	Yes	No	Yes	Yes	Yes	No	No	Yes
BeiDou	Yes	No	Yes	No	Yes	No	No	No
SBAS	Yes	Yes	Yes	Yes	Yes	Yes	Yes	Yes
Number of channels	220	136	220	220	220	226	120	120
Horizontal accuracy in static surveying	2.5 mm + 0.5 mm/km	2 mm + 0.5 mm/km	2.5 mm + 0.5 mm/km	5 mm + 0.5 mm/km	3 mm + 0.5 mm/km	4 mm + 0.5 mm/km	3 mm + 1 mm/km	3 mm + 1 mm/km
Vertical accuracy in static surveying	5 mm + 1 mm/km	5 mm + 1 mm/km	5 mm + 1 mm/km	5 mm + 1 mm/km	3.5 mm + 1 mm/km	3.5 mm + 1 mm/km	3.5 mm + 1 mm/km	6 mm + 1 mm/km
Horizontal accuracy in RTK surveying	10 mm + 1 mm/km	10 mm + 1 mm/km	10 mm + 1 mm/km	10 mm + 1 mm/km	8 mm + 1 mm/km	8 mm + 1 mm/km	10 mm + 1 mm/km	10 mm + 1 mm/km
Vertical accuracy in RTK surveying	20 mm + 1 mm/km	20 mm + 1 mm/km	20 mm + 1 mm/km	20 mm + 1 mm/km	15 mm + 1 mm/km	15 mm + 1 mm/km	20 mm + 1 mm/km	20 mm + 1 mm/km

(continued)

TABLE 6.4 (*Continued*)
Comparison of Geodetic GNSS Receivers with Smart Antenna

Specifications	UniStrong G970	Altus APS-3L	Acnovo GX9	South S82-V	Trimble R4-3	Topcon Hiper V	Geomax Zenith 25	Leica Viva GS15
Internal memory	256 MB (SD card)	1 GB (SD card)	64 MB	65 MB	11 MB		Micro SD (8 GB)	SD up to 32 GB
Water/dustproof	IP67	IP67	IP67	IP67	IP67	IP67	IP68	IP67
Measurement update	50 Hz	25 Hz	50 Hz	1 Hz	10 Hz	1 Hz	20 Hz	20 Hz
Battery	Li-Ion	2 × Li-Ion	Li-Ion	Li-Ion	Li-Ion	Li-Ion	Li-Ion	2 × Li-Ion
Bluetooth	2.00 + EDR	2.00 + EDR	2.00 + EDR	2.00 + EDR	2.00 + EDR	2.00 + EDR	2.00 + EDR	2.00 + EDR
Wi-Fi	No	No	No	No	No	No	No	No
UHF radio	410–470 MHz	406–470 MHz	410–470 MHz	410–470 MHz	450	Embedded	Embedded	Embedded
GSM module	Embedded	Embedded	Embedded	Embedded	Embedded	Embedded	Embedded	Embedded
Operating temperature	From –40° to +75°C	From –40° to +65°C	From –40° to +60°C	From –40° to +60°C	From –40° to +70°C	From –40° to +65°C	From –30° to +60°C	From –45° to +65°C
Weight	1.2 kg	1.3 kg	1.2 kg		1.52 kg	1.32 kg	1.2 kg	1.34 kg
Dimension	184 × 96 mm	178 × 89.7 mm		184 × 96 mm	190 × 102 mm	190 × 109 mm	95 × 198 mm	

6.6.1 Description of the Various Applications

GNSS-aided navigation and positioning is used in many sectors of the economy, as well as in science, technology, tourism, research, and surveying. GNSS can be utilized wherever precise three-dimensional positional data has a significant role to play. A few important sectors are detailed next.

6.6.1.1 Location-Based Services (LBS)

Location-based services (LBS) are services based on the current position of a user (e.g., mobile communications network users equipped with a cellphone). Normally, the mobile station (e.g., cellphone) must be logged on and its position given in order to request or obtain specific information/services from the provider. An example of this is the distribution of local information, such as the location of the nearest restaurant or automatically providing the caller position to emergency number services (E-911 or E-112).

The prerequisite for LBS is the determination of accurate position information. Location is determined either through signals from the cellphone network or by using satellite signals.

The location of the user is given either with absolute geographic coordinates (longitude and latitude) or relative to the position of a given reference point (e.g., the user is located within a radius of 500 m to the monument). There are basically two kinds of services provided, known as "push services" and "pull services." A push service sends the user information on the basis of his or her position without their having to request it (e.g., "In the vicinity is ..."). A pull service requires that the user first request the information from the service (e.g., calling an emergency number E-911 or E-112).

Knowing the location is of critical importance for surviving emergencies. However, public security and rescue services have shown in a study that 60% of those making emergency calls with mobile telephones were unable to communicate their exact position (in comparison to 2% of callers from fixed-net telephones). Every year within the European Union there are 80 million emergency calls made, of these 50% are made with mobile telephones.

The determination of the user's position can either be obtained within the mobile station or by the mobile network. For determining the position, the mobile station refers to information from the mobile communication network or satellite signals.

Countless technologies for positioning have already been introduced and have been standardized. Few of these are currently being used and it remains to be seen if all the ideas will ever be realized. In Europe, the most common applications currently being used are

- Position determination through the identification of active cells in the cellphone network (Cell-ID). This procedure is also known as Cell of Origin (COO) or Cell Global Identity (CGI).
- Position determination by the time delay of GSM-Signals TA (Timing Advance). TA is a parameter in GSM networks through which the distance to the base station can be determined.
- Satellite positioning through satellite navigation, for example, GNSS.

6.6.1.2 Commerce and Industry

For the time being, road transportation continues to be the biggest market for GNSS. Of a total market value estimated at $60 billion in 2005, $21.6 billion alone was accounted for road transportation and $10.6 billion for telecommunications technology. Vehicles will be equipped with a computer and a screen, so that a suitable map showing the position can be displayed at all times. This will enable selecting the best route to the destination. During traffic jams, alternative routes can be easily determined and the computer will calculate the journey time and the amount of fuel needed to get there.

Vehicle navigation systems will direct the driver to his or her destination with visual and audible directions and recommendations. Using the necessary maps stored on CD-ROM and position estimates based on GNSS, the system will determine the most favorable routes.

GNSS is already used as a matter of course in conventional navigation (aviation and shipping). Many trains are equipped with GNSS receivers that relay the train's position to stations down the line. This enables personnel to inform passengers of the arrival time of a train.

GNSS can be used for locating vehicles or as an antitheft device. Armored cars, limousines, and trucks carrying valuable or hazardous cargo will be fitted with GNSS. An alarm will automatically be set off if the vehicle deviates from its prescribed route. With the press of a button, the driver can also operate the alarm. Antitheft devices will be equipped with GNSS receivers, allowing the vehicle to be electronically immobilized as soon as monitoring centers receive a signal.

GNSS can assist in emergency calls. This concept has already been developed to the marketing level. An automobile equipped with an onboard GNSS receiver connected to a crash detector. In the event of an accident, the car signals an emergency call center providing precise information about the direction the vehicle was traveling and its current location. As a result, the consequences of an accident can be made less severe and other drivers can be given advanced warning.

Railways are another highly critical transportation applications where human life is dependent on technology functioning correctly. Here, precautions need to be taken against system failures. This is typically achieved through backup systems, where the same task is performed in parallel by redundant equipment. During ideal operating situations, independent sources provide identical information. Well-devised systems indicate (in addition to a standard warning message) if the available data is insufficiently reliable. If this is the case, the system can switch to another sensor as its primary data source, providing self-monitoring and correction. GNSS can provide a vital role here in improving system reliability and safety.

Other possible uses for GNSS include:

- Navigation systems
- Fleet management
- Geographical tachographs
- Railways
- Transport companies, logistics in general (aircraft, waterborne, craft and road vehicles)

- Automatic container movements
- Extensive storage sites
- Laying pipelines (geodesy in general)
- Positioning of drill platforms
- Development of open-pit mining
- Reclamation of landfill sites
- Exploration of geological deposits

6.6.1.3 Communications Technology

Synchronizing computer clocks is vital in situations with separated processors. The foundation of this is a highly accurate reference clock used to receive GNSS satellite signals along with Network Time Protocol (NTP), specified in RFC 1305. Other possible uses for GNSS include:

- Synchronization of system time-staggered message transfer
- Synchronization in common frequency radio networks

6.6.1.4 Agriculture and Forestry

GNSS contributes to precision farming in the form of area and use management, and the mapping of sites in terms of yield potential. In a precision farming system, combined harvest yields are recorded by GNSS and processed initially into specific plots on digital maps. Soil samples are located with the help of GNSS and the data added to the system. Analysis of these entries then serves to establish the amount of fertilizer that needs to be applied to each point. The application maps are converted into a form that onboard computers can process and are transferred to these computer using memory cards. In this way, optimal practices can be devised over a long term that can provide high time/resource savings and environmental conservation.

Other possible uses for GNSS include:

- Use and planning of areas
- Monitoring of fallow land
- Planning and managing of crop rotation
- Use of harvesting equipment
- Seeding and spreading fertilizer
- Optimizing logging operations
- Pest management
- Mapping diseased and infested areas

There are many conceivable GNSS applications for the forest industry as well. The U.S. Department of Agriculture (USDA) Forest Service GPS Steering Committee 1992 has identified over 130 possible applications in this field. Examples of some of these applications are briefly detailed next:

- Optimizing log transportation—By equipping commercial vehicle fleets with onboard computers and GNSS, and using remote data transfer facilities, transport vehicles can be efficiently directed from central operations units.

- Inventory management—Manual identification prior to timber harvesting is made redundant by satellite navigation. For the workers on site, GNSS can be used as a tool for carrying out specific instructions.
- Soil conservation—By using GNSS, remote roads and tracks used in harvesting wood can be identified and their frequency of use established.
- Management of private woodlots—In wooded areas divided into small parcels, cost-effective and highly mechanized harvesting processes can be employed using GNSS, allowing the transport of increased quantities of timber.

6.6.1.5 Science and Research

With the advent of the use of aerial and satellite imaging in archeology, GNSS has also become firmly established in this field. By combining GIS (Geographic Information Systems) with satellite and aerial photography, as well as GNSS and 3D modeling, it has been possible to answer some of the following questions:

- What conclusions regarding the distribution of cultures can be made based on the location of the finds?
- Is there a correlation between areas favoring the growth of certain arable plants and the spread of certain cultures?
- What did the landscape look like in this vicinity 2000 years ago?

Surveyors use (D)GPS to quickly and efficiently carry out surveys (satellite geodesy) to within an accuracy of a millimeter. For surveyors, the introduction of satellite-based surveying represents a progress comparable to that between the abacus and the computer. The applications are endless. They range from land registry and property surveys to surveying roads, railway lines, rivers, and the ocean depths. Geological variations and deformations can be measured and landslides and other potential catastrophes can be monitored.

In land surveying, GNSS has virtually become the exclusive method for pinpointing sites in basic grids. Everywhere around the world, continental and national GNSS networks are developing that, in conjunction with the global International Terrestrial Reference Frame (ITRF), provide consistent and highly accurate networks of points for density and point-to-point measurements. At a regional level, the number of tenders to set up GNSS networks as a basis for geoinformation systems and cadastral land surveys is growing.

GNSS already has an established place in photogrammetry. Apart from determining coordinates for ground reference points, GNSS is regularly used to determine aerial survey navigation and camera coordinates for aerotriangulation. Using this method, over 90% of ground reference points can be dispensed with. Future reconnaissance satellites will be equipped with GNSS receivers to aid the evaluation of data for producing and updating maps in underdeveloped countries.

In hydrography, GNSS can be used to determine the exact height of a survey boat. This can simplify the establishment of clearly defined reference points. The expectation is that usable GNSS procedures in this field will be operational in the near future.

Other possible areas of application for GNSS are

- Archeology
- Seismology (geophysics)
- Glaciology (geophysics)
- Geology (mapping)
- Surveying deposits (mineralogy, geology)
- Physics (flow measurements, time standardization measurement)
- Scientific expeditions
- Engineering sciences (e.g., shipbuilding, general construction industry)
- Cartography
- Geography
- Geoinformation technology
- Forestry and agricultural sciences
- Landscape ecology
- Geodesy
- Aerospace sciences

6.6.1.6 Tourism and Sport

In sailplane and hang glider competitions, GNSS receivers are often used to maintain protocols with no risk of bribery. Also, GNSS can be used to locate persons who have found themselves in a maritime or alpine emergency (i.e., search and rescue). Other possible uses for GNSS include:

- Route planning and selecting points of particular significance (natural and culturally/historically significant monuments)
- Orienteering (training routes)
- Outdoor activities and trekking
- Sporting activities

6.6.1.7 Military

GNSS is used anywhere where combatants, vehicles, aircrafts, and guided missiles are deployed in unfamiliar terrain. GNSS is also suitable for marking the position of minefields and underground depots, as it enables a location to be determined and found again without any great difficulty. As a rule, the more accurate, encrypted GNSS signal (PPS) is used for military applications and can only be used by authorized agencies.

6.6.1.8 Time Measurement

GNSS provides the opportunity to exactly measure time on a global basis. Around the world, "time" (UTC) can be accurately determined to within 1 and 60 ns. Measuring time with GNSS is much more accurate than with so-called radio clocks, which are unable to compensate for signal travel times between the transmitter and the receiver. If, for example, the receiver is 300 km from the radio clock transmitter, the signal travel time already accounts for 1 ms, which is 10,000 times less accurate than time measured by a GNSS receiver. Globally precise time measurements are necessary for synchronizing control and communications facilities, for example.

Currently, the most common method for making precision time comparisons between clocks in different places is a "common-view" comparison with the help of GNSS satellites. Institutes that wish to compare clocks measure the same GNSS satellite signals at the same time and calculate the time difference between the local clocks and GNSS system time. As a result of differences in the measurement, the difference between the clocks at the two institutes can be determined. Because this involves a differential process, GNSS clock status is irrelevant. Time comparisons between the PTB and time institutes are made in this way throughout the world. The PTB atomic clock status, determined with the help of GNSS, is also relayed to the International Bureau for Weights and Measures (BIPM) in Paris for calculating the international atomic time scales UTC and TAI (International Atomic Time).

BIBLIOGRAPHY

Global positioning system standard positioning service signal specification. https://www.navcen.uscg.gov/pubs/gps/sigspec/gpssps1.pdf

Hofmann-Wellenhof, B., H. Lichtenegger, and J. Collins. 2001. *GPS: Theory and Practice*. Wien: Springer Verlag.

Jean-Marie, Z. 2009. GPS: Essentials of satellite navigation. *Compendium: Theory and Principles of Satellite Navigation*, Overview of GPS/GNSS Systems and Applications, Thalwil, Switzerland: U-Blox.

Kaplan, E. D. 1996. *Understanding GPS: Principles and Applications*. Boston: Artech House.

Parkinson, B. W., P. Enge, P. Axelrad, and J. J. Spilker Jr., eds. 1996. *Global Positioning System: Theory and Applications*, Vol. 2. Washington, DC: American Institute of Aeronautics and Astronautics.

7 Obtaining and Processing Unmanned Aerial Vehicle (UAV) Data for Surveying and Mapping

7.1 OBTAINING SURVEYING DATA USING SMALL UNMANNED AERIAL VEHICLES (UAVs)

Since the first flight of the Wright Brothers on December 17, 1903, at 10:35 a.m. all the airplanes have been piloted by humans. As flying machines developed, more skilled personnel were required to handle them. A new era that started with the first aircraft that flew purely on turbojet power, the Henkel He178, led to new achievements but also required pilots with better knowledge of those machines (Blom 2010).

As the complexity of flying vehicles increased year by year, it became clearer that the weaker part of the flying machine and human equation was the human. The discussion about a pilotless airplane started during World War I and the first flying machine flew a little bit after, sometime around 1916 using AM Low Radio Control techniques. World War II pushed aircraft design technology further and along with it creating even more reliable electronics something that benefited of course also unmanned aircrafts.

As the years passed and with the side developments in many scientific areas, designers were able to create lighter and more reliable flying machines. The invention of transistors in 1947 lead the electronics industry to new frontiers, and unmanned aerial vehicles became "smarter" and faster. It was clear to anyone that the future belongs to unmanned aerial vehicles (UAVs) since those flying machines were not putting humans life at risk.

Flying a machine from a remote control center with minimum human stress, and with "on the fly" consultation of other technicians and scientists interpreting the results during the flight, it was pure superiority. The advantages were almost endless and that led the U.S. Air Force (USAF) in 2011 to train more UAV pilots. And that year became one more historical point for the aviation: human pilots became less necessary than remote machine pilots (Wolverton 2012).

A UAV—also commonly known as a drone, unmanned aircraft system (UAS), and by several other names—is an aircraft without a human pilot aboard. Typically, UAVs are presented by fixed wing, rotary, kites, balloons, gliders, and other aerial platforms. Samples of these different platform types are given in Figure 7.1. Given the opportunity to carry image acquisition and other sensor platforms (Paiva 2011), the UAV has been called the next surveying game changer.

FIGURE 7.1 Different surveying UAV platforms.

Unmanned flight technology is known from 1940s, nowadays we may observe a boost in its development. Especially the idea and technology spread out again when we started to watch at the TV those great Hawk and Predators UAVs to participate to strategic missions.

UAVs can be classified by a broad number of characteristics, including weight, engine, and payload (Valavanis et al. 2008).

Important aspects for performance are the following:

- Weight
- Endurance and range
- Maximum altitude
- Wing loading
- Engine type
- Power/thrust loading

All these parameters are parts of the aerodynamic analysis during UAV designing. An important classification (Valavanis et al. 2008) of the UAV by the total weight is given in Table 7.1.

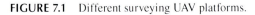

TABLE 7.1
UAV Classification by Weight

Designation	Weight Range (kg)
Super heavy	>2000
Heavy	200–2000
Medium	50–200
Lightweight	5–50
Micro	<5

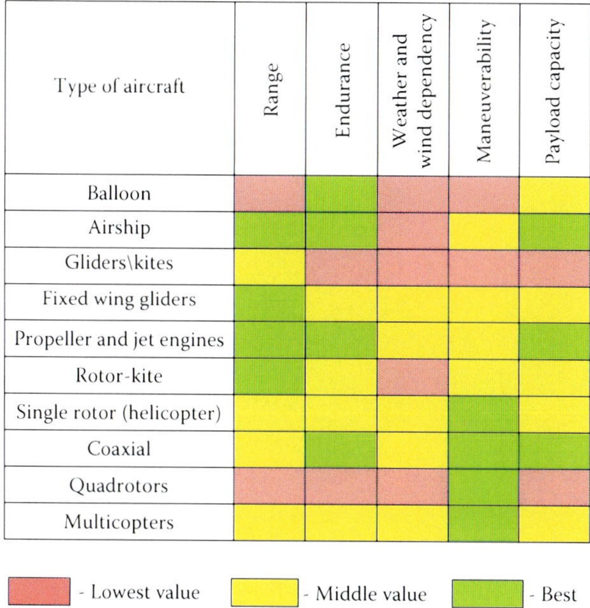

Type of aircraft
Range
Endurance
Weather and wind dependency
Maneuverability
Payload capacity
Balloon
Airship
Gliders\kites
Fixed wing gliders
Propeller and jet engines
Rotor-kite
Single rotor (helicopter)
Coaxial
Quadrotors
Multicopters
- Lowest value
- Middle value
- Best

FIGURE 7.2 Pros and cons of the different UAV types.

Another type of classification is based on endurance:

- There are high endurance UAVs capable for staying airborne more than 24 hours and for range more than 1500 km.
- Medium endurance UAVs are capable for flying between 5 and 24 hours and for a range between 100 and 400 km.
- Low endurance UAVs fly for less than 5 hours and for a range of less than 100 km.

Of course depending on the type of use and manufacturer or even nation, there are classifications by payload, primary use, altitude, and so forth. This chapter will make use of a low altitude, low endurance UAV. Comparative characteristics of the different UAV platforms in terms of the aforementioned classification parameters (reworked from Eisenbeiss 2011) are given in Figure 7.2.

The U.S. Department of Defense (Army UAS CoE Staff 2016) uses an integrated classification based on weight, altitude, and speed, which is commonly accepted among UAV hobbyists and professionals (Table 7.2). From Table 7.3 groups 1 and 2 are of the most use in surveying engineering and geospatial mapping application scenarios.

The most important toolset that enables mapping, surveillance and remote sensing is a sensor suite. Typically, a sensor suite combines sensors, antennas, power supply, processors, and transmitters. Sensors are determined based on the part of the electromagnetic waves spectrum (depicted in Figure 7.3) to which they are sensitive.

It is visible from Figure 7.3 that all the sensors that potentially can be deployed in the Earth Observing System (EOS) may only collect data in spectral diapasons that

TABLE 7.2
UAV Classification

Category	Size	Maximum Cross Takeoff Weight (lb)	Normal Operating Altitude (ft)	Airspeed (knots)
Group 1	Small	0–20	<1200 AGL	<100
Group 2	Medium	21–55	<3500 AGL	<250
Group 3	Large	<1320	<18,000 MSL	<250
Group 4	Larger	>1320	<18,000 MSL	Any airspeed
Group 5	Largest	>1320	>1800	Any airspeed

Note: AGL is above ground level and MSL is mean sea level.

are not absorbed by the atmosphere called "atmospheric windows." Typical sensors in major UAV payloads include electrooptical (EO), infrared (IR), light to voltage (LTV), synthetic aperture radar (SAR), infrared SAR, and ground moving target indicator (GMTI). Obviously for surveying and mapping applications, EO sensors are of primary interest. There are two different types of EO sensors: CMOS (complementary metal-oxide semiconductors) and CCD (charged-couple devices). Both are a kind of 2D—a matrix—formed by photodiodes (Sandau 2009).

Both systems are starting from exactly the same point: They must convert photons to electrons. After the conversion we need to read the accumulated charge value. For the CCD this is happening by traveling the charge to the edge of the array. For the CMOS every pixel that is forming the active matrix has several transistors that are amplifying and moving the charge to their destinations. This approach is more flexible because each pixel can be read individually.

There are some noticeable differences between CCD and CMOS systems that affect photographs (Ohta 2007):

- CCD sensors create high-quality, low-noise images. CMOS sensors, traditionally, are more susceptible to noise.
- Because each pixel on a CMOS sensor has several transistors located next to it, the light sensitivity of a CMOS chip tends to be lower. Many of the photons hitting the chip hit the transistors instead of the photodiode.
- CMOS traditionally consumes little power. Implementing a sensor in CMOS yields a low-power sensor.
- CCDs use a process that consumes lots of power. CCDs consume as much as 100 times more power than an equivalent CMOS sensor.
- CMOS chips can be fabricated on just about any standard silicon production line, so they tend to be extremely inexpensive compared to CCD sensors.
- CCD sensors have been mass-produced for a longer period of time, so they are more mature. They tend to have higher quality and more pixels.

Technical characteristics of the most typical surveying UAV sensor suites are summarized (Colomina and Molina 2014) in Table 7.3.

TABLE 7.3
Typical UAV Sensors Parameters

Vendor/ Model	Sensor Type[a]	Format Type or Scanning Pattern[b]	Resolution (MPx)	Size (mm^2)	Pixel Size (lm)	Weight (kg)	Frame Rate (fps)	Speed[c] (Hz)	Spectral Range	Spectral Bands/ Resolution (nm)	Thermal Sensitivity (μK)	Range (m)	Angular Resolution (deg)	FOV (deg)	Laser Class and *k* (nm)	Frequency (kp/s)
DJI/Phantom FC300	EO	SF	CMOS			0.2		80–8000 (fp)								
Phase One/iXA 180	EO	MF	CCD 80	53.7 × 40.4	5.2	1.7	0.7	4000 (fp)								
Trimble/IQ180	EO	MF	CCD 80	53.7 × 40.4	5.2	1.5		1600 (ls)								
Hasselblad/ H4D-60	EO	MF	CCD 60	53.7 × 40.2	6	1.8	0.7	1000 (ls)								
Sony/NEX-7	EO	SF	CMOS	23.5 × 15.6	3.9	0.35	2.3	800 (ls)								
Ricoh/ GXRA16	EO	SF	CMOS	24.3 × 15.6	4.8	0.35	3.0	4000 (fp)								
Tetracam/ MiniMCA-6	MS		CMOS 1.3	6.66 × 5.34	5.2 × 5.2	0.7			450–1050 (nm)							
Quest Innovations/ Condor-5 VVN-285	MS		CCD 1.4	10.2 × 8.3	7.5 × 8.1	0.8			400–1000 (nm)							
Rikola Ltd./ Hyperspectral Camera	HS		CMOS	5.6 × 5.6	5.5	0.6			500–900 (nm)	40/10						
Headwall Photonics/ Micro-Hyperspec X-series NIR	HS		InGaAs	9.6 × 9.6	30	1.025			900–1700 (nm)	62/12.9						

(Continued)

TABLE 7.3 (*Continued*)
Typical UAV Sensors Parameters

Vendor/ Model	Sensor Type[a]	Format Type or Scanning Pattern[b]	Resolution (MPx)	Size (mm²)	Pixel Size (lm)	Weight (kg)	Frame Rate (fps)	Speed[c] (Hz)	Spectral Range	Spectral Bands/ Resolution (nm)	Thermal Sensitivity (μK)	Range (m)	Angular Resolution (deg)	FOV (deg)	Laser Class and *k* (nm)	Frequency (kp/s)
DRS Technologies/ Tamarisc 320	T		640 × 480		17	0.5		9–30 (fp)								
FLIR/TAU 2 640	T		640 × 512	10.8 × 8.7	17	0.07			7.5–13.5 (lm)		≤50					
Thermoteknix Systems Ltd./ Miricle 307K-25	T		640 × 480	16 × 12.8	25	0.105			8–12 (lm)		≤50					
ibeo Automotive Systems/ IBEO LUX	L	4 scanning parallel lines				1						100	(H) 0.125 (V) 0.8	(H) 110 (V) 3.2	Class A 905	22
Velodyne/ HDL-32E	L	32 laser/ detector pairs				2						200	(H) – (V) 1.33	(H) 360 (V) 41	Class A 905	700
RIEGL/ VQ-820-GU	L	1 Scanning Line				4						1000	(H) 0.01 (V) N/A	(H) 60 (V) N/A	Class 3B 532	200

[a] Sensor type: EO, electrooptical (visible); MS, multispectral; HS, hyperspectral; T, thermal; L, lidar.
[b] Format type or scanning pattern: MF, medium format; SF, small format.
[c] Speed (Hz): fp, focal plane shutter; ls, leaf shutter.

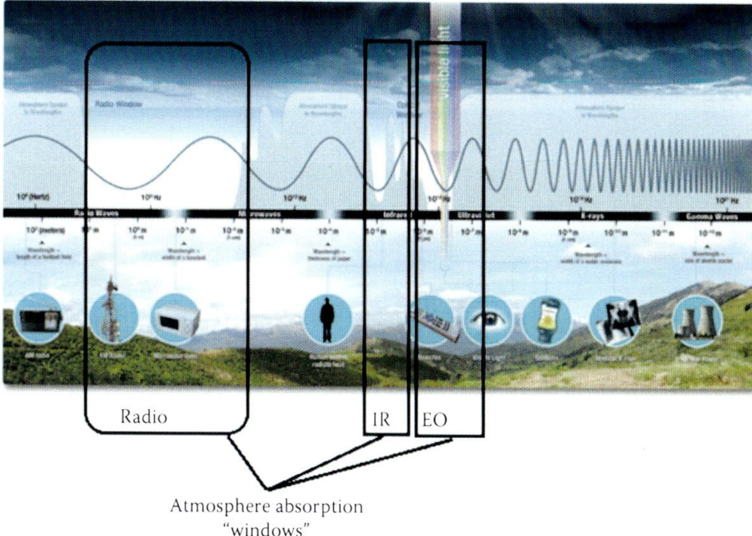

FIGURE 7.3 Electromagnetic waves spectrum diagram. (Modified from National Aeronautics and Space Administration, Science Mission Directorate. 2010. Introduction to the electromagnetic spectrum. http://missionscience.nasa.gov/ems/01_intro.html.)

Nowadays research and development efforts comprise many reports on successful deploying of lidar sensors in small UAV sensor suites (Lin et al. 2011; Tulldahl and Larsson 2014). Deploying of IfSAR radar sensors (Goshi et al. 2012; Koo et al. 2012) for 3D modeling is also a promising direction, since accuracy of that method is similar to the geometric leveling described in Chapter 2. Figure 7.3 depicts samples of the small UAV sensor suites integrated for the surveying in mapping purposes. Figure 7.4 depicts sensors installed on Michigan Technological University UAV platforms.

FIGURE 7.4 Sensors installed at Michigan Tech in UAV (Canon Reibel EOS) and E38 small UAV (Canon SX 260 HS).

The next section is devoted to considerations of small UAV use for surveying and mapping.

7.1.1 Principles of UAV Use for Surveying and Mapping Projects

The demand on various services needed for Earth surface surveying in the geospatial market is quite high. Therefore, the pace of information collection and processing should be at an appropriate rate. Typically, these processes are local in nature, affecting small areas scattered in space, and are quite numerous. The concentration of changes occurring on land within the administrative and territorial units is not large. Therefore, use of manned aircraft, albeit small, will result in unnecessary costs. From an economic standpoint it becomes clearly unjustified. Even if there is a justification of this expense related to the urgency and importance of the task (such as emergency management situations), these projects are not seldom absent of favorable weather conditions, leading to the disruption of the project schedule. Use of ground-only data collection methods and land surveys also increases the cost of work. In these described circumstances, the use of easy to transport, remotely controlled aircrafts equipped with the necessary imaging equipment allows very quick performance of all the necessary project stages and allows for flexible adjustments to the ongoing technology, even in the absence of stable weather conditions needed for the traditional aerial survey ("window" is always there). Also, UAV imagery can be deployed for the generation of high-resolution digital surface models (DSM) and can alternate or complement lidar point clouds.

Luhmann et al. (2006) introduced a categorization scheme of measurement techniques that trades sizes of the measured objects versus required accuracy. Modification of this scheme is presented in Figure 7.5.

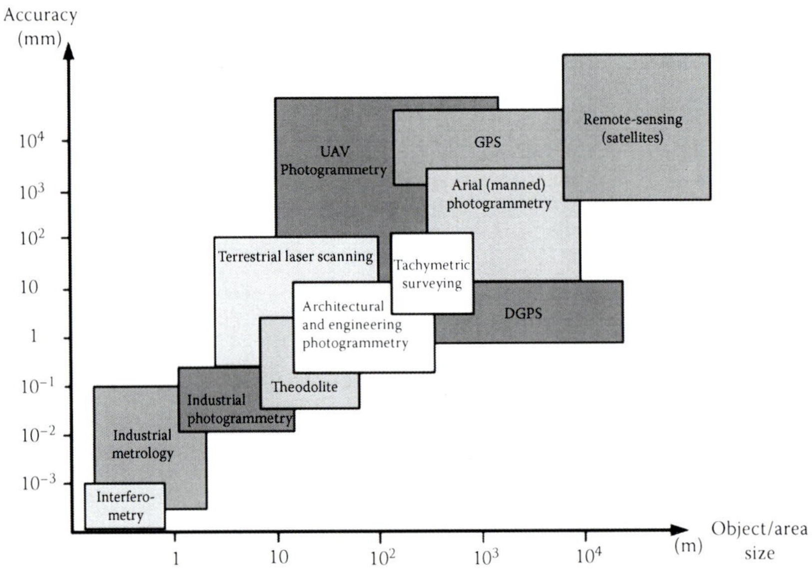

FIGURE 7.5 Relationship between project size and accuracy of the surveying methods.

As for UAV deployment limitations, it is necessary to recognize that especially small UAVs are limited in size and resolution of sensors that could be cared by such platforms. Usually that small format sensors are not calibrated and are less stable then standard large format aerial cameras. Given the small footprint on the ground of UAV imagery, it is necessary to collect more images to compare to traditional aerial survey. Not all state-of-the-art photogrammetric software systems support UAV imagery processing. A review of some toolsets is given in Section 7.3.

As for aerial traffic safety issues, small UAVs are not equipped with air traffic communication equipment and collision avoidance systems, compare to manned aircrafts (Colomina et al. 2008). Therefore, due to the lack of communication with air traffic authorities, small UAVs are restricted to flight in line-of-sight (LOS) and to operate with a backup pilot. Besides UAV control stations communication frequencies (typically 30–40 MHz) signals may interfere with another civil segment device (remote controlled cars and hobbyist aircrafts), therefore there is the possibility of signal jamming.

7.2 UAV PHOTOGRAMMETRY

Photogrammetrically there are two possible types of UAV aerial imaging: (1) vertical images taken inside a ±3° angle from a UAV and (2) oblique images that are taken with a more than 3° angle (Kraus 2007). The oblique images are also divided to high oblique images, which show a part of the horizon, and low oblique, which strictly show the terrain.

The problem with the oblique imagery is mainly scale. They do not have uniform scale and is it difficult to apply metric photogrammetry. There are solutions for this problem, but for surveying engineering applications vertical or near-vertical images are essential (Paine and Kiser 2012).

7.2.1 Image Scale

The most important type of images for our research are vertical images where the scale is known by the following equation (Falkner and Morgan 2002):

$$S = \frac{f}{H - Ha} \tag{7.1}$$

where f is the focal length, H is the flying altitude above ground, and Ha is the average altitude of the area above sea level. By knowing the focal length (it is a part of the camera) that we are using, and the flying altitude (known by the GPS device), we can calculate the scale. If we know the scale, then we can measure the size of objects or features directly over the imagery.

7.2.2 Ground Coverage (Footprint) by Single Image

The next step defining UAV imagery metrics is to find out how much space each image is covering on the ground.

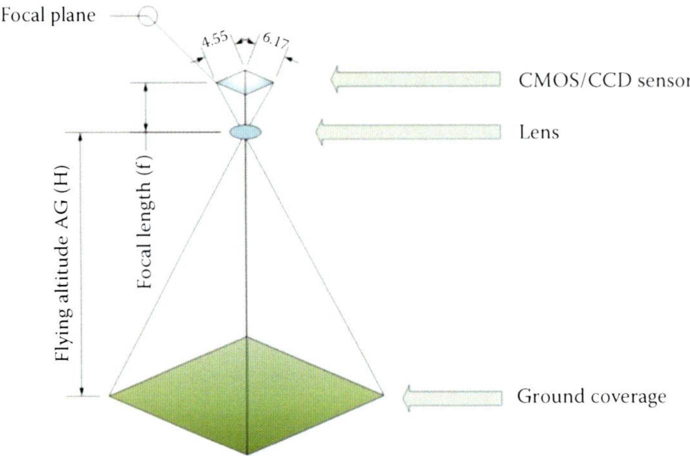

FIGURE 7.6 Basic geometry for ground coverage.

At the side of an exposure station (defined by spatial position of image principal point at the moment of image acquisition) we may observe the film frame size or the CMOS/CCD sensor size as it is depicted in Figure 7.6.

The formula for calculating the ground coverage per image taken is the following:

$$\frac{f}{\text{Altitude}} = \frac{Cx}{Gx} \tag{7.2}$$

where f is the principal focal distance, altitude is of course the flying distance above ground, Cx is the side of the CMOS/CCD matrix, and Gx is the corresponding ground side (Falkner and Morgan 2002).

For example, we are applying dependency for the Canon SX 260 HS camera: focal length = 4.5 mm with flying, altitude = 108 m, and $Cx = 6.17$ mm:

$$\frac{f}{\text{Altitude}} = \frac{Cx}{Gx} \Rightarrow \frac{0.0045}{108} = \frac{0.00617}{Gx} \Rightarrow Gx = 148.08 \text{ m}$$

Applying the same formula for the other side/CCD:

$$\frac{f}{\text{Altitude}} = \frac{Cx}{Gx} \Rightarrow \frac{0.0045}{108} = \frac{0.00455}{Gy} \Rightarrow Gy = 109.2$$

It means that every single image acquired by the UAV flying at 108 m height is covering a rectangle of 148 × 109 m (the fractions here are adding nothing to the accuracy, so we prefer dealing with integers than floating-point numbers). The covering area is 148 × 109 = 16,132 m^2.

7.2.3 Ground Sample Distance

The ground sample distance, or GSD, is one of the most important parts of any imagery specification. It is given for every single pixel of the image, that is, how much ground is covered (Linder 2003).

The way to calculate this number is similar to the ground coverage but here we are also using the CMOS or CCD resolution. A very important point here is that we are using the theoretical maximum number of pixels that the CMOS/CCD matrix is providing and not the effective number. This is happening because we are looking at each pixel coverage and not the effective performance of the matrix.

For example, the sensor area (CMOS) for Cannon SX 260 HS camera is 6.17 × 4.55 mm. We must convert these numbers to microns, that is, 6.17×10^6 μm and 4.55×10^6 μm, and therefore the sensor area is 28.0735×10^6 μm.

The total number of pixels for the CMOS is 12,800,000, which is equal to 12.8×10^6 pixels. Dividing the sensor area over the total number of pixels we get the pixel area:

$$\text{Pixel Area} = \frac{28.0735 * 10^6}{12.8 * 10^6} = 2.193421875\ \mu\text{m}^2$$

Since the pixels are squares, then the pixel side is

$$\sqrt{\text{Pixel Area}} = \sqrt{2.193421875} = 1.48\ \mu\text{m}$$

For example, the specific scale for the UAV mission case considered in Section 7.2.2, we have

$$\text{GSD} = \text{Scale} * \text{Pixel Area} = \frac{108}{0.045} * 1.4\ \mu\text{m} * 10^{-6} = 0.03552\ \text{m} = 3.552\ \text{cm}$$

Every single pixel of the Cannon SX 260 HS camera, while flying on a UAV at 108 m above ground, is covering 3.552 cm on the ground.

7.2.3.1 Ground Sample Distance: Comparative Analysis

A higher GSD means better resolution and also means clearer identification of objects on the ground. Nowadays, one of the highest resolution imagery satellites, GeoEye-1 of Digital Globe, is providing GSD of 50 cm for image swathes of 15.2 km. It is easy here to recognize the advantages and disadvantages of using satellite image acquisition methods compare to UAVs. We are getting by far better resolution but for a limited ground extent. For covering the same area of a single satellite image, we should plan somewhat more than 20 missions with the UAV. Still, this is feasible task that can be implemented during a one-day field project.

7.2.4 Multiview Processing

One of the great benefits that photogrammetry and computer vision are deploying is the stereoscopic/multiview processing capability. It was inspired by the ability of the human eye–brain system to form a 3D model of the scene by perception of slightly different views of objects by the left and right eye. Specifically, during stereo perception provided by the phenomena, known as a human stereopsis, the human eyes function similar to extremely complex optical "cameras" that are observing images of an object from slightly different positions. This position is the distance between the eyes. Our brain then is interpreting those images giving the sense of depth.

The stereoscopic view is doing exactly the same. Using stereoscopic glasses, or even more advanced devices, we are looking at two consecutive images that are taken during the flight mission with these glasses. The left image with our left eye and the right image with the right eye. Even if it sounds difficult, the stereoscopic glasses are doing this task extremely easy. As soon as we form this type of view the images are turning to a 3D view and it is allowing us to perform even more calculations. Besides the metric part of the stereoscopic view, another very important aspect is the interpretation of the images since the 3D perspective is adding to the total benefit and it is giving the real sense of the terrain.

But for forming the stereoscopic view there are some parameters that need to be matched with the photographs. The end lap and the side lap are the most important aspects of imagery that should be matched. The minimum end lap, which is the overlap of two consecutive images taken at the direction of flight path, must be at least 60%. The side lap, which is the overlap of two images that are taken during the flight in parallel lines, should be at least 30% (Edward et al. 2001).

Since we have the ground coverage by a single image, then it is easy to calculate the end lap and side lap. For the example considered in previous sections, we will have the following UAV image acquisition configuration as it is depicted in Figures 7.7 and 7.8.

$$\text{Distance between points 1 and 2} = 148\text{ m} - (148\text{ m} \times 0.6) = 59.2\text{ m}$$

That is, the end lap = 59.2 m. The side lap is

$$109\text{ m} - (109 \times 0.3) = 76.3\text{ m}$$

The interpretation of these numbers is the following: During UAV flight, every 59.2 m must acquire an image. The distance between the flying strip lines is 76.3 m.

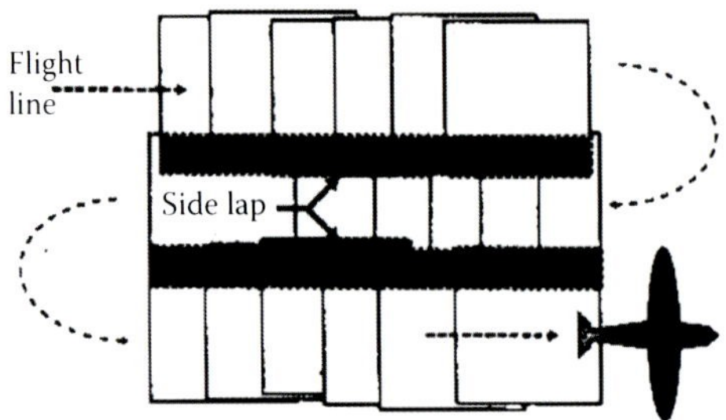

FIGURE 7.7 End lap and side lap for aerial imaging mission.

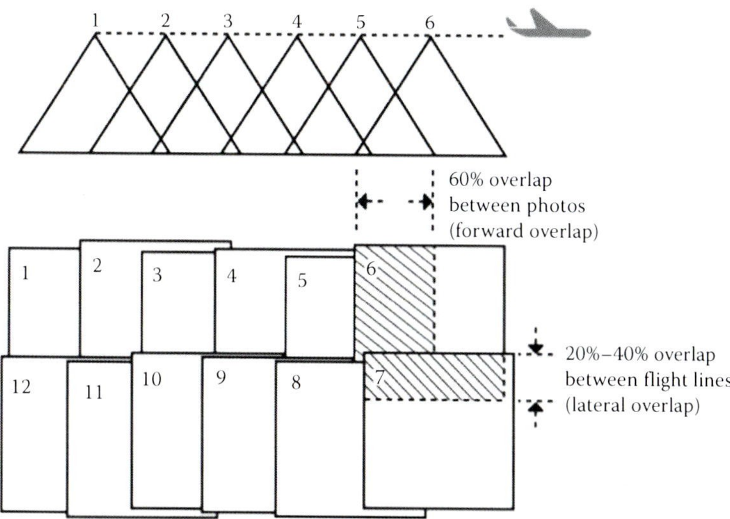

FIGURE 7.8 End lap and exposure station position.

These were the basic calculations for starting to plan a mission for acquiring aerial images. There are more tasks that need to be done and those tasks are very crucial for the success.

7.2.5 Photogrammetric Computer Vision: Influence on UAV Imaging Technologies

UAV image processing mainly consists of two stages: (1) imagery orientation based on onboard telemetric (GPS/IMU [inertial measurement unit] combination) data or ground control and (2) feature extraction and digital surface model generation. Nowadays developments in computer vision are widely applied in software packages that are used on everyday basis for that UAV data processing. One of the pillars that is deployed by both traditional photogrammetry and computer vision is a bundle adjustment technology. Triggs et al. (2000) introduce the schematic history of the bundle adjustment algorithms development as shown in Figure 7.9.

Novel approaches that came from computer vision side include structure from motion (SFM) and scale-invariant feature transform (SIFT) algorithms deployed by many modern UAV processing toolsets.

Classical photogrammetric processing is based on concepts of deploying calibrated cameras and two well-known major concepts—collinearity and coplanarity—which forms photogrammetric equations of the same name. Both concepts are illustrated in Figure 7.10.

The SFM approach deploys the idea of simultaneous processing of multiple images obtained from different locations, as depicted in Figure 7.11.

Comparative analysis of the classical photogrammetry versus SFM approaches is given in Table 7.4. It is visible from the table that SFM has inheritance of the classical

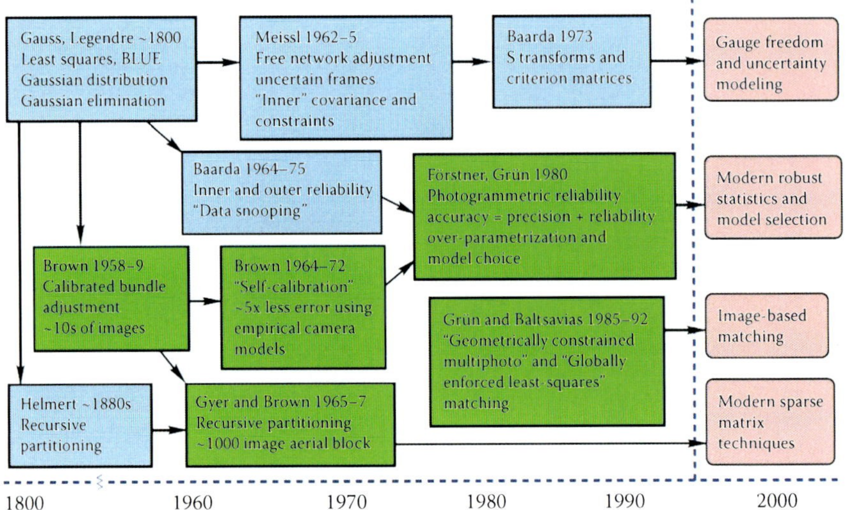

FIGURE 7.9 Bundle adjustment algorithms (historical perspective).

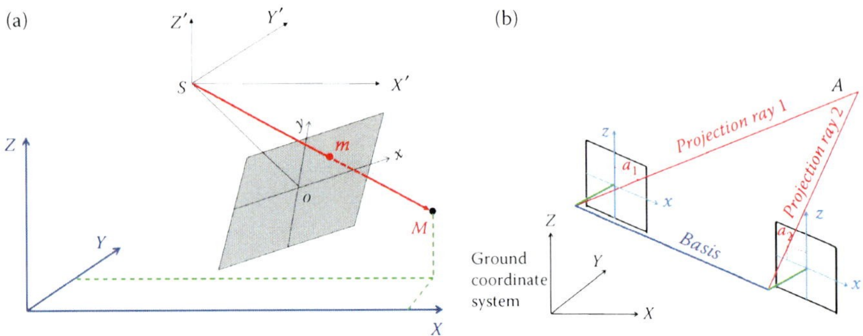

FIGURE 7.10 Classical photogrammetry. (a) Concept of collinearity: projection center, image of terrain point, and point of terrain belong to the same straight line. (b) Concept of coplanarity: stereopair basis, left and right (conjugate) projection rays, and terrain point belong to the same plane.

photogrammetry fundamental principles and brings some novel elements that became possible toward resent extensive developments in computer sciences. Specifically, most of the SFM-based systems are designed to perform the following fully automated workflows: reconstruct scene geometry and camera motion from two or more images; generating high-resolution textured 3D polygonal models; and implementing camera (auto)calibration. That workflow is enabled by the following image processing and computer vision algorithms: feature detection, image matching, camera autocalibration, distortion compensation, 3D point cloud generation, and mesh gridding.

One of the most promising techniques applied in the frame of SFM is SIFT, which deploys the idea of the pixel-based matching of features detected on images obtained

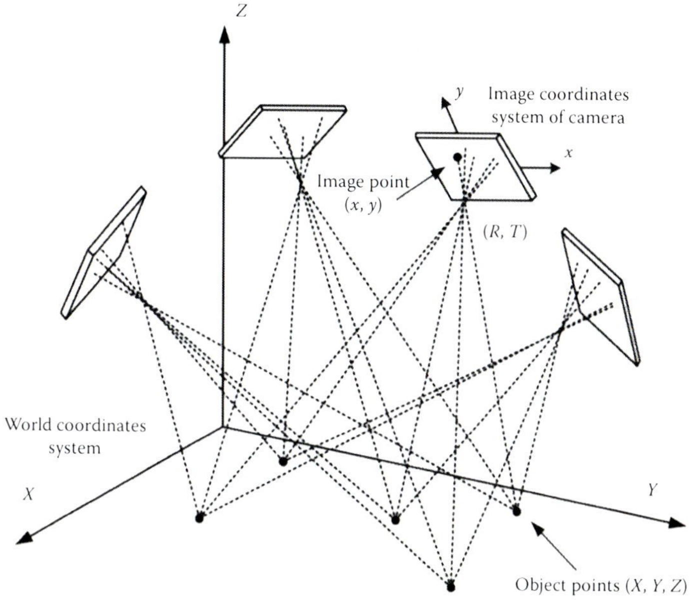

FIGURE 7.11 Structure from motion concept illustration.

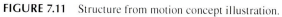

TABLE 7.4
Classical Photogrammetry versus Structure from Motion

Approach/ Solution	Classical Photogrammetry	Structure from Motion
Image/Scene Orientation	1. Two views of five distinct 3D points (Kruppa 1913) 2. Relative position and orientation of cameras 3. Spatial positions of points	1. Two views of eight distinct 3D points (Longuet-Higgins 1984; Hartley and Sturm 1997) 2. Relative position and orientation of cameras 3. Spatial positions of points 4. Processing with camera calibration parameters
3D models form planar images generation	1. Exploit prior knowledge about the scene to reduce ambiguity and ensure optimal geometrical conditions 2. Overlapping area from two to six images 3. Specific conditions for image acquisition 4. Ensure optimal geometric conditions and guarantee robust mathematical solutions	1. Use of excessive number of corresponding image points in multiple views 2. The more pictures the better; redundancy is a crucial factor for robust solution 3. No strict geometrical requirements for image acquisition 4. No explicit knowledge about the object/ scene 5. No selective process for point identification

in a crowd-source way by different sensors and from various locations. Based on Lowe (2004) the SIFT principle can be illustrated as shown in Figure 7.12.

After performing matching refining by SIFT, a traditional SFM algorithm is deployed in a sequence that is described in Table 7.4.

SFM processing for the UAV and terrestrial imagery is usually comprised of the following technological steps: topological editing (gapes), refining models (decimate mesh, smooth), merging submodels, crops processing (background removal), georeferencing, orthorectification, and export results to generic file formats (ply, dxf, obj, pdf, etc.).

State-of-the-art software systems that are deployed for UAV imagery processing are listed in Table 7.5.

With the aforementioned photogrammetric computer vision capabilities and technology, 3D vision technology may in the future even dominate lidar due to simplicity and the high level of automation of that tool (Nouwakpo et al. 2015). Imaging-based results have the following advantages over lidar instruments: subpixel accuracy of the algorithms, higher point densities, and fully automated operational workflows.

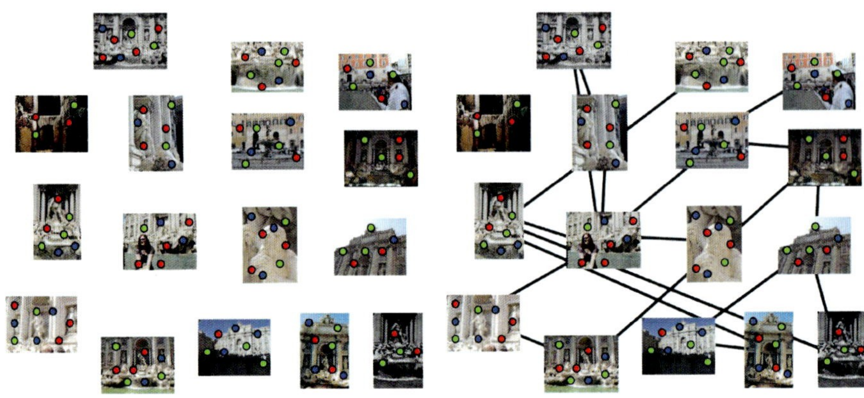

FIGURE 7.12 SIFT algorithm principles: feature detection on multiple images and then feature matching for each pair of images. (Agarwal, S. et al. 2010. *Computer* 43, no. 6:40–47.)

TABLE 7.5
State-of-the-Art Software Toolsets Deployed for UAV Data Processing

UAV Specific Packages	Close-Range Packages	Open-Source Packages
AgiSoft PhotoScan	PhotoModeller	Visual SFM
Pix4D	iWitness	Bundler
AeroGIS	DLR SGM	PMVS/CMVS
SimActive	RMA DSM	
Adam	IGN MicMac	
OrthoVista	Joanneum Research	
Icaros		
Mensic software		

For the most of data samples processed for this book we deployed the AgiSoft PhotoScan system. Specific processing steps and results are described in the following sections.

7.3 UAV MAPPING PROJECT MANAGEMENT CONSIDERATIONS

This section contains practical samples, considerations, and data from a UAV mapping project performed at Michigan Technological University deploying the E38 small UAV and Cannon SX 260 HS camera.

7.3.1 Basic Tasks

A successful UAV image acquisition mission has three important tasks that need to be completed in a very specific way: (1) target area and preparation for the mission; (2) specification for the camera that will be deployed in the mission; and (3) specification for the UAV flight parameters. Each one of these tasks is critical for the mission. Each one needs to be completed before moving to the next, because each tasks informs the next one (Falkner and Morgan 2002).

7.3.1.1 Target Area and Preparation

It is mandatory to collect information about the area of the UAV mission. The operational workflow associated with that activity can be termed a target area circle (Levin et al. 2013) and is depicted in Figure 7.13. We start with visual inspection of the area. We are looking for features or objects that could possibly interfere with the imagery or with the flying mission itself. Maybe there are high buildings near the site or high trees.

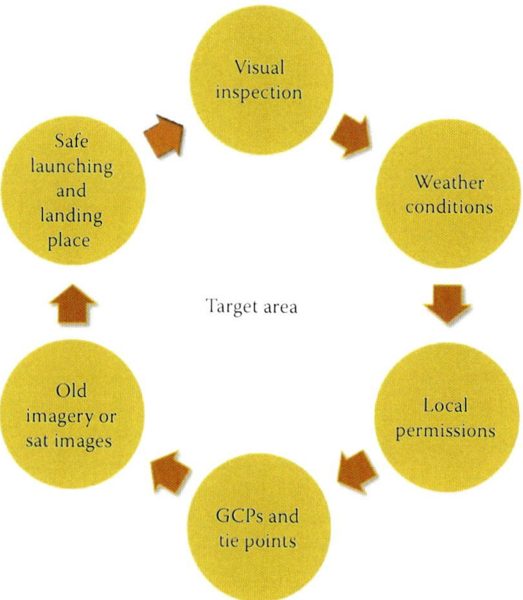

FIGURE 7.13 The target area circle.

Next is to check the weather conditions of the area. On the Internet there are weather portals that keep historical data. We are looking for weather conditions around the period that we will deploy the UAV.

The next step is to obtain any local permission from authorities. A visit to the county or the police station will help avoid any problems.

Now is the time for marking the ground control points (or GCPs), sometimes called tie points. Tie points are very obvious features that we are able to recognize in at least two consecutive images. For example, building corners, crossroads, and statues. We use these points for stitching images together and creating a photo mosaic. When covering large areas we need multiple images. By stitching two, three, or more images we are creating a bigger synthetic image that is representing the whole area. This synthetic image (synthetic because is not only one image but many more together) is called a photo mosaic.

GCPs are ground points that we have measured the coordinates of by using surveying methods. Theoretically, with just three points we can have acceptable accuracy, but practically six points allow the ability to apply special models such as polynomial models for photogrammetric analysis if this is required.

The GCPs should be spread out covering the whole target area. On the ground we are marking these points with such a way that would be visible on the imagery while the UAV passes over the area to perform the mission. But how we can design those marking and how big should they be? A typical ground marker has a cross shape with some specific dimensions that are related with the mission characteristics (Aber et al. 2010). For example, we assume that we have GSD = 5 cm. According to Section 7.2.3, this is also the pixel size (see Figure 7.14).

A typical ground marker (considering a GSD of 5 cm) will have a shape and size as shown in Figure 7.15.

Another practical suggestion is that every marker we are using should be installed in such a way to be oriented indicating the north. This helps during the image interpretation. Figure 7.16 depicts practical samples of the ground control points setup at Michigan Tech.

The next step is to try to find old imagery of the target area. This will help us to identify possible problems of the area. The last part is to find a safe place for launching and landing the UAV.

7.3.1.2 Camera Specifications

The camera we used is a part of the E38 UAV kit. Its characteristics were described in Section 7.2.2.

7.3.1.3 UAV Flight Parameters Specifications

7.3.1.3.1 UAV Speed

The term *recycling time* refers to how fast a camera is able to shoot an image, save the image on the storage card, then shoot the next image. This time for our camera is 2.7 s minimum. Since pushing the electronics of the camera along with other components of the UAV is never a good idea, we are introducing a safety factor of 50% to increase the recycling time to be sure we give the camera the

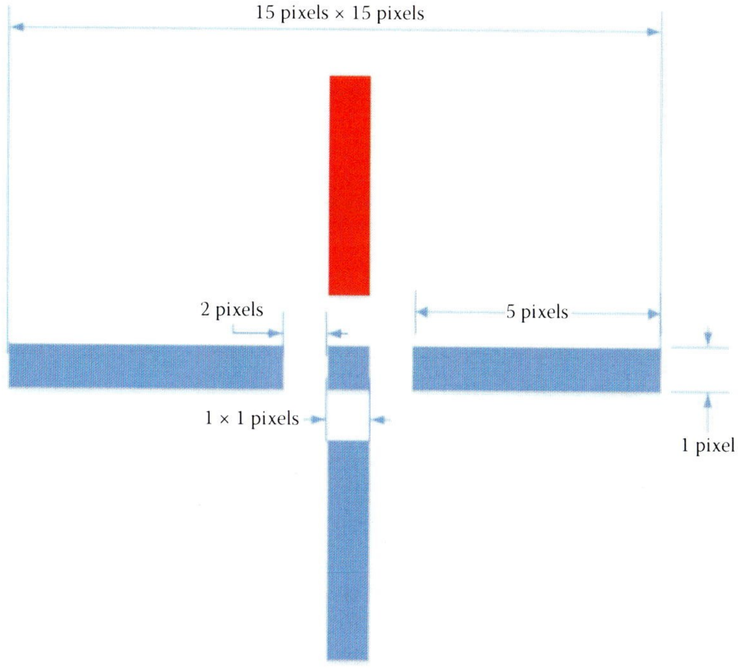

FIGURE 7.14 Ground marker distances by pixels.

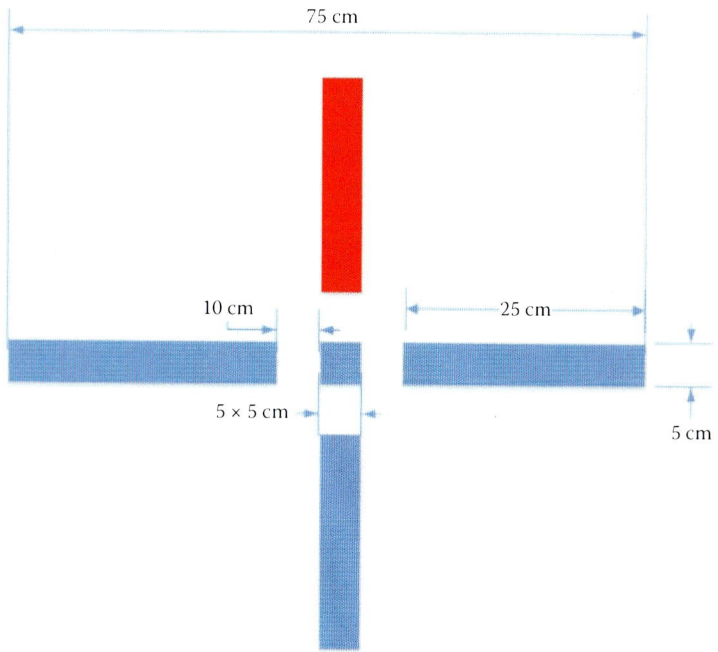

FIGURE 7.15 Ground marker final dimensions.

FIGURE 7.16 Ground control points. (a) GPS measurement process, (b) GCP installed on the ground, and (c) GCP on UAV image.

necessary time to complete a full imaging procedure cycle (Levin et al. 2013). Therefore, we have

$$2.7\,\text{s} \times 1.5\ (\text{safety factor }50\%) = 4.05\,\text{s}.$$

Having the distance of 59.2 m that must separate each image (because of the end lap) and the recycling time of the camera, we calculate the UAV mission speed as

$$\text{Mission Speed} = \frac{59.2\,\text{m}}{4.05\,\text{s}} = 14.617\,\text{m/s}$$

7.3.2 Mission Planning Sample

After performing the basic aforementioned calculations, it is time to see how we can plan a real mission. Through this knowledge will also investigate other factors that could affect the mission such as the weather conditions and especially the atmospheric conditions. We have to start as we previously illustrated by scouting the area.

7.3.2.1 Project Area

The project area is a rectangular area near the Michigan Tech campus and is better known with the local name of Massie's Strawberry Fields (Figure 7.17). The corners of the marked area have the following UTM coordinates:

- 5210695.68N 382012.66E
- 5210692.34N 382233.93E

The total area is 250 m × 200 m = 50,000 m^2 or 0.05 km^2. The area is partially residential but is mostly a kind of junkyard for abandon vehicles.

7.3.2.2 Mission Planner

Mission Planner is an open-source software for the autopilot embedded system (Ardu Pilot 2016). It is used by several map vendors, including Google, Microsoft,

FIGURE 7.17 Massie's field UAV target area.

and OpenStreetMap for base maps. Mission Planner is capable of creating way-points and upload those waypoints into our embedded system, as depicted in Figure 7.18.

Keeping in mind the basic calculations we made earlier, we are setting the auto-pilot software based on the following: altitude, 108 m; end lap, 59 m (60% overlap); and side lap, 76 m (30% overlap).

FIGURE 7.18 Mission planner GUI.

The software will create a flight path for our mission and will also provide UTM coordinates for the waypoints where the exposure station (camera) should be activated.

7.3.2.3 The Flight Path

The following table is how a flight path looks. There are several numbers but the most important here are the coordinates for matching the navigation path and the exposure station trigger point. For example, the rows starting with 0, 1, and 2 contain flight path waypoints and the altitude, but row 3 is the activating the exposure station and that is why the software marks it with altitude 0 (column 10):

0	1	0	16	0	0	0	0	47.039722	-88.553817	190.520004	1
1	0	3	16	0.000000	0.000000	0.000000	0.000000	47.039249	-88.552346	108.000000	1
2	0	3	16	0.000000	0.000000	0.000000	0.000000	47.039052	-88.552346	108.000000	1
3	0	3	203	0.000000	0.000000	0.000000	0.000000	0.000000	0.000000	0.000000	1
4	0	3	16	0.000000	0.000000	0.000000	0.000000	47.038827	-88.552346	108.000000	1

Additionally, to this event (altitude zero) the embedded system is taking specific control signal indicating that the exposure station trigger is set to "on" mode.

The screenshot in Figure 7.19 presents the corresponding waypoints exactly as they are on the software. As mentioned earlier, we are able to isolate from the waypoints table the exposure station points where the camera should be triggered on. According to our mission planner the points are as shown in Figure 7.20.

Besides the exposure station points here we are also getting the images footprint (how much area each image is covering). We have to remind here that each image is covering a ground area of 148 m × 109 m and we are following a 60% end lap and 30% side lap.

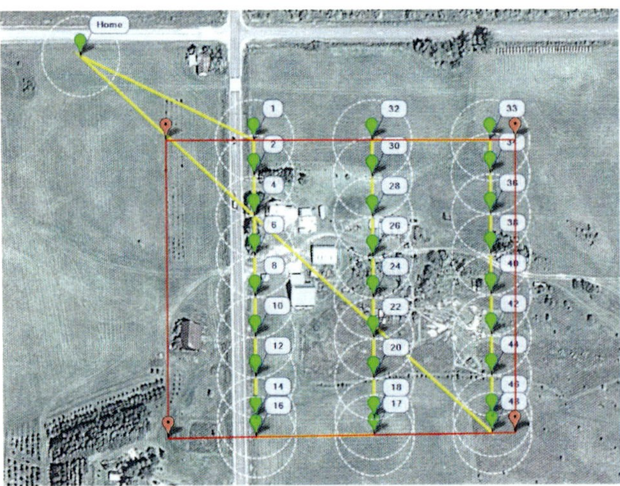

FIGURE 7.19 Mission planner, navigation waypoints.

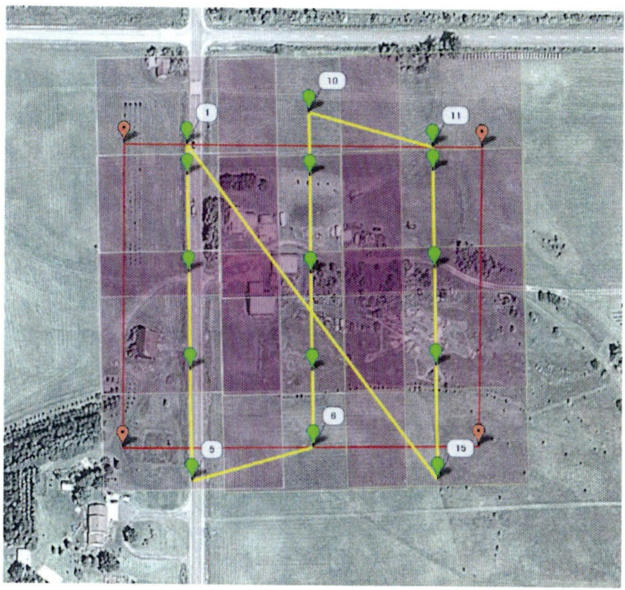

FIGURE 7.20 Mission planner, images footprint.

7.3.2.4 Mission Results

During the mission (we flew three times testing flying parameters and different camera setups) we acquired more than 200 images.

At this point we have to remind ourselves we didn't use any ground control points because our purpose was to establish the base theory for deploying the sensor and not to apply metric photogrammetry.

On the other hand, we have had plenty of tie points since the area is almost full of infrastructure and objects. From these images we select 10–12 images and proceed to create a photo mosaic of the area. The image is making use of a spherical effect representation according to the camera distortion profile, as is shown in Figure 7.21. The final product is in Figure 7.22. Please note that

- We are covering more than the specified area (because of the imagery footprint).
- For forming this wide file of the main area we used four images.
- The size of the actual file is 90 Mb and the resolution is around 9000 × 3000 pxls.
- The GSD for the product is 3.5 cm.

7.3.3 Results Analysis

First, we were using a camera with a very wide lens and focal length of 4.5 mm. The acquired products with such a camera are always good candidates for big distortions (Wolf and Dewitt 2000), especially when the image angle is not perfectly vertical but near vertical. Remember that near vertical angle imagery means ±3°.

FIGURE 7.21 Spherical representation of the target area.

FIGURE 7.22 Final result.

Also the camera is mounted in a stable way on the fuselage and not with a mechanism that will help the camera to compensate small angles. A mechanism like this would also be helping with absorbing some vibrations. Here, because of the mounting, all the vibrations are passing through the camera body and to the lens.

The camera, because of the fast shutter speed necessary for acquiring "stable" images, introduces some noise at the UAV imagery-based products such as orthophotos and 3D models. This is also a result of using cheap electronics, sensitive to electromagnetic interferences (Sandau 2009). Using cheap electronics could also lead to problems with the heat that is formed inside the fuselage. The body of the camera is constructed of aluminum and this conducts heat. If the mission is more than 15–20 minutes, then we should probably use a cooling solution for the camera and for the embedded systems of the autopilot.

7.3.4 Factors Affecting the Mission

There are several factors that contribute one way or another to the success or failure of an image acquisition mission. The most important are the lighting and the atmospheric conditions. Both as subjects for analysis are huge topics, but here we will illustrate only the most important parts.

7.3.4.1 Atmospheric Conditions

An aerial imaging acquisition mission is considered as successful only when one is getting the quality of images that serve the scope of the planning mission. But the most common feature that every image must have is to be illuminated by as much light as possible. Poor atmospheric conditions such as bad weather, crosswinds, or wrong season might heavily affect the results, and by this we mean extensive hidden areas around the objects or the introduction of image noise, or blurry images (Aber et al. 2010).

The position of the sun is the main factor creating shadows. Typically, the best time for aerial image acquisition is between 10 a.m. and 2 p.m. Depending by the latitude of the area that we are proceeding with the imaging, somewhere around 12 p.m. is the best time because of the minimum shadows as it is depicted in Figure 7.23.

Another important issue that we must avoid is the phenomenon called "hot spot" (see Figure 7.24). This is a relatively small area—a spot on the image—that heavily reflects the sunlight. Normally hot spots are created because of the direct alignment of the camera and the sunlight. The hot spot is located at the antisolar point that is on the ground point that is opposite the sun in relatively with the camera. Hot spots have been subject to great investigation in recent years. A fast explanation of the phenomenon is the absence of visible shadows causes the spot to display more luminosity than the surrounding area.

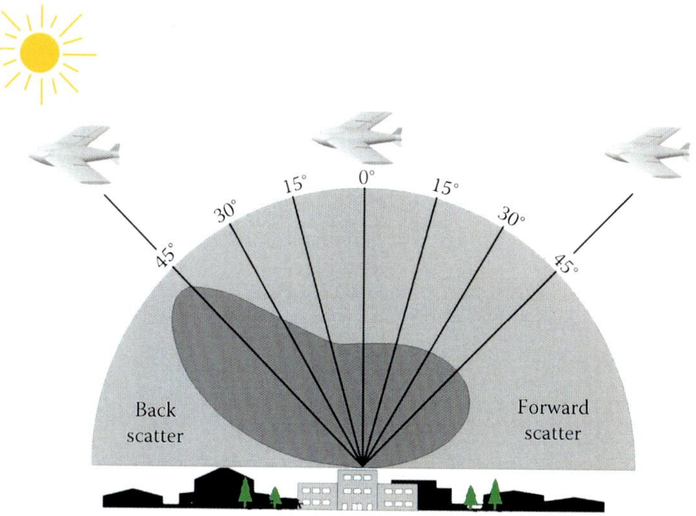

FIGURE 7.23 Diagram of aerial photography and typical BRDF. Reflectivity pattern in the solar plane is marked here by the black area. Maximum reflectivity occurs directly back toward the sun. (Adapted from Weishampel, J. F. et al. 1994. *Remote Sensing of Environment* 47, no. 2:120–131.)

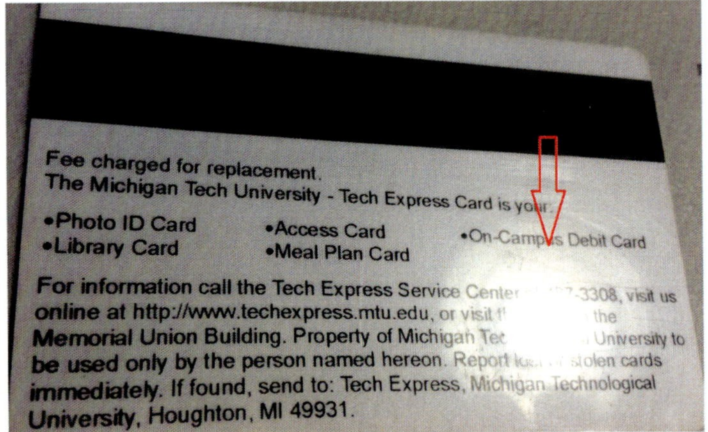

FIGURE 7.24 Sample of hot spot on image obtained from smartphone camera.

7.3.4.2 Weather

The weather is probably the most dangerous condition that could badly affect the mission. Most electronics are vulnerable to humidity or, even worse, the direct exposure to water due to rain or snow.

Enclosing the electronics in a tightly sealed waterproof box does not solve the problem because of the accumulated heat inside the case. This is a very dangerous condition that could easily lead to a massive failure. On the other hand, the GPS units must be positioned in a way that should be able to have clear sight of the sky for better results.

Most UAVs have small holes or hatches on their fuselage to allow a small amount of air to pass inside the fuselage to cool the engine and other electronics, like the battery regulators or the Arduino boards. Flying on a rainy day is a high-risk situation for allowing water to pass inside the airplane and change the balance of the aircraft or damage the electronics.

7.3.4.2.1 High Temperature

High environmental temperatures are extremely dangerous for electronics that are working on a small airplane bay like the UAV fuselage. Additionally, the camera also has problems working in high temperatures (Sandau 2009). The adoption of small holes on the fuselage that could allow air intake to cool down the components during hot weather, but could compromise the mission on rainy days.

A solution could be small fans positioned on the back part of the UAV to remove hot air from inside the fuselage. This solution could be also good for the camera electronics because it lowers the temperature at the sensor bay where the camera is located. It should be very good design and aerodynamic analysis because it is possible the fan's electric engine could produce noise at the images.

7.3.4.2.2 Wind

Winds are the main reason for having to repeat a flight mission. The main problems caused by winds are drift, crab, drift and crab, and oblique images when we need vertical (Paine and Kiser 2012). Drift is the result of the UAV being unable to keep the planned navigation bearing. The crab is the result keeping the planned navigation bearing but because of side winds the UAV does not have an alignment yaw position with the bearing. Because of extensive crosswinds there is a great possibility that we will have a combination of the two situations, drift and crab (see Figure 7.25).

Oblique images are a result of the UAV not being able to compensate for winds from multiple directions or circulations due to the presence of big buildings near the site of the mission, or trees or high slopes, and during the navigation reposition moves, the exposure station goes off. As we mentioned before, vertical and near-vertical images are those that are taken with at angle between 0° and ±3°. As mentioned earlier, oblique images might not always be a catastrophe. They give better results for image interpretation since the interpreter is looking at the area in a more familiar perspective than the vertical.

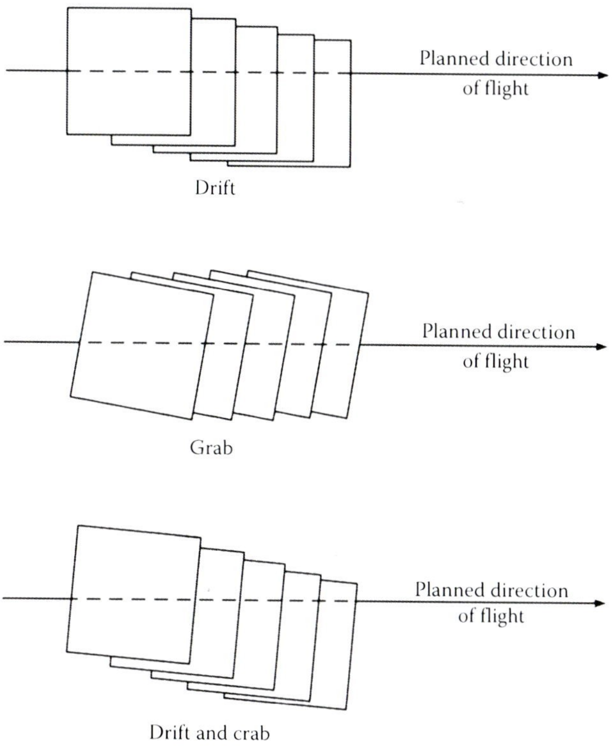

FIGURE 7.25 Drift, crab and drift and crab visual representation. (Adapted from Paine, D. P. and Kiser, J. D. 2012. *Aerial Photography and Image Interpretation*. Hoboken, NJ: John Wiley & Sons.)

7.3.4.2.3 Season

The season should be considered before executing a mission for aerial imaging. The answer depends on the purpose of the image acquisition. If we need to analyze the terrain of a rural area or in a place with extensive vegetation, then the best time is fall or spring when the trees are losing their leaves. This could prove to be a bit tricky since many type of vegetation are losing their leaves in a very short period before snow or other bad weather, maybe just few days before. And the situation could be even worse because when trees are losing their leaves, a windy day could bring them down very fast, and in a couple of days there may also be the first snowfall.

If you need to analyze more populated places like towns, you usually do not have to deal with extensive vegetation. Although hot spots can occur, summer does provide a good time for high-resolution images.

Winter is also a good period for an image acquisition mission if you really need to follow vegetation or animal immigration patterns.

7.4 UAV MAPPING PROJECT COST CONSIDERATIONS

This book would be incomplete with an analysis of the cost for adopting UAV technology for image acquisition missions. First, we will analyze today's prices for the most important equipment components.

7.4.1 Camera and Component Prices

A compact camera with 12 MPxls resolution and a CMOS of 6.17 mm × 4.55xx with focal length of 4.5 mm and Recycling Time of 2700 ms costs $200.

A DSLR camera with a 16 MPxls resolution, CMOS of 22.4 mm × 16.7 mm with fixed lens of 35 mm focal length and recycling time of 3000 ms costs $700.

Memory cards like SD cards (SDHC format) with fast accessing time (to reduce the bottleneck of writing data) with a capacity of 32GB costs around $60.

The camera stabilization mechanism is something that you typically cannot buy from any store. It is something that you have to design and construct. The design could take two to three days and the construction another two days. The cost of the materials could be less than $100 but one week of engineering design and manufacturing could be analyzed as following: The U.S. Bureau of Labor Statistics for 2016 gives the median price per hour for a mechanical engineer at $38.74. For one week's labor (including the construction) the total cost is $1550. Additionally, the material cost is another $100 for a total cost of $1650.

7.4.2 The UAV

There are many different types of UAVs available. The most interesting for aerial image acquisition missions are the fixed-wing UAVs since they are capable of covering extensive areas.

The cheapest UAV that one can buy is just $300; it is only the airplane almost ready to fly, with the necessary openings for the camera. You just have to buy the camera separately. For this type of UAV, the camera is mounted in a stable way at the fuselage, so there is no way to use a stabilization mechanism. The images of this type of UAV are "suffering" from the acquisition of a large amount of nonvertical images, so if you have to use this type of UAV (with stable mount of the camera), then you have to expect a lot of "garbage" imagery and multiple attempts to acquire high-quality images (Levin et al. 2013).

There are also UAVs that come ready to use for imaging missions. They include a camera already mounted (again in a stable way). One only has to charge the batteries and fly. These types of UAV have a minimum cost of $1200. As mentioned earlier, using these types UAVs you should expect to collect a big amount of "garbage" images. But by flying multiple missions, especially in areas with poor atmospheric conditions, at the end of the day you will have enough quality images.

The last type of UAVs is those that are completely created for a purpose. Designing a UAV like this has multiple benefits, but the cost is a disadvantage. Mechanical engineers with knowledge of how to design a UAV should be involved (Lozano 2010). The project should also involve electrical engineers, computer engineers, and geospatial engineers. It is really a multidisciplinary job. Of course, the total cost of a UAV would be much more than all the others mentioned earlier. The work from the team that created the first version of a photogrammetric small UAV was estimated to be $2500 but in fact ended up being $30,000.

Designing the fuselage and selecting the proper airfoil is a 10-day job for one person. Splitting the work between two people can split the time in half. Electronics and wiring for the UAV is a three-day job. The manufacturing takes 10 days, but this is an outsourced job that costs, according to multiple offers, around $2000. The stabilization mechanism of the camera, plus all the necessary components brings the project to a one-month timetable and a cost of $10,000 (assuming the engineers are not exclusively working on this project because there is much "dead time" between the component fabrication and assembly). Adding also the cost of engines, switches, the camera, and so forth, the total cost comes close to $12,000.

7.4.3 The Mission's Cost

In the example that we analyzed in Section 7.2.2, the total area that we covered was 50,000 m^2. The distance, according to the software, that the UAV traveled is 1192.8 m. With the speed of 14.67 m/s, the time to acquire the imagery was 81.3 s (round up to 82 s). The preparation between flights is no more than 10 min. We have to change the batteries and to make a visual check that the UAV is in good condition. Nothing else.

We were able to finish the mission with just one flight (we flew three missions just because we also needed to perform some flying tests), but assuming that we needed a second one, we add another 82 s. The total time of image acquisition is 164 s, that is 2 min 44 s, plus the time of preparation 10 min at the beginning plus another 10 min between the missions, that is 20 min. The total time is 23 min.

In our example we did not use any ground control points, but if we were using them we should add the labor for one surveyor. According to the Bureau of Labor Statistics the mean per hour cost of a surveyor is $27. For a total number of three GCPs using the real-time-kinematic (RTK) method, the work is less than 1 hour, so the labor cost is $27.

The total time of the mission is also around two hours (GCPs plus flights), so we have for our example one mechanical engineer for two hours, that is $39 × 2 = $78, plus $54 for the surveyor (who should be there for the total amount of the time), for a total labor amount for the mission of $132.

Using photogrammetric software, we rectified the images and created the photomosaic of the whole area in 20 minutes. If we need to create an orthophoto, we would first create a DTM of the area and then proceed with the Orthophoto. That could be one more hour of desktop work.

The total time for this mission would be two hours office work for the surveyor at a cost of $54, plus $132 for the field work for a total of $186. Since there could be things that go wrong or other expenses unaccounted for, such as travel costs from the field to the office (i.e., gas), we need to assess another $50. The total cost is $236 for four hours of work from two people.

7.4.4 Small UAV (Hobbyist Level) Mapping Flight Recommendations

Our experience indicates that specifically for light, small UAVs reaching waypoints circle is different in practice then in theory. As soon as the autopilot hits that area, point is marked as completed, and is loading the next point on the list. The autopilot does not search for the waypoint position but for the area. In response we created a new flight pattern where the waypoints sequence is organized such that after completion even strip SUAV is following to the next even strip instead of consequent odd strip. That novel flight pattern sequence is depicted in Figure 7.26. It is visible from Figure 7.26 that instead of following 1-2-3-4-5-6-7-8 sequence SUAV adopted flight plan was designed as 1-3-5-7-2-4-6-8 sequence.

The pattern we are using here provides adequate space for the UAV to turn and match the navigation points, but mostly is designed to facilitate the navigation through winds of more than 7 m/s. Our practical experiments indicated feasibility and sufficient image quality of that nonstandard flight pattern.

7.5 PROFESSIONAL-LEVEL SUAV PROJECT RECOMMENDATIONS

There are multiple professional-level UAV systems available on the surveying instruments market. Most of the largest vendors of geospatial equipment are developing their own proprietary lines of UAS solutions. For example, Trimble's line of products includes the ZX5 hexacopter and UX5 (UX5HP) fixed-wing systems for photogrammetric quality image processing (Trimble UAS, n.d.). Leica Geosystems has the AIBOT-X6 UAS multicopter instrument that is accompanied with the Aibotix AiProFlight flight planning software package (Leica Geosystems 2016). Topcon's Falcon-8 UAS has both fixed wing and rotary wing hardware with specialized flight planning and image processing software packages (Topcon 2015). And RIEGL is pioneering a UAV-based lidar scanning with its VUX-1 lightweight sensor (RIEGL UAV 2016).

In the current section, the most critical aspects of the surveying data acquisition and processing projects will be considered based on the Trimble UX5 HP SUAS system (Trimble 2013).

(a)

(b)

FIGURE 7.26 UAV flight plans samples. (a) Traditional and (b) small UAV adopted flight pattern.

The UX5 HP aerial imaging rover follows a preprogrammed path where takeoff, flight, and landing require minimal human intervention. If required, the crew operating the aircraft from the ground can intervene to change the flight path or to land. In some cases, such as communications failure or loss of GPS signal, a preprogrammed intervention is automatically activated to reestablish signal loss or to terminate the flight early and complete a safe landing.

The UX5 HP aerial imaging rover holds a camera that takes aerial images over the defined area. During the flight, all pictures are acquired at a specified height, along parallel lines with specified overlap between the image exposures. At the same time, precision GNSS-based position information is recorded during a postprocessing kinematic (PPK) survey to achieve highly accurate position information for the captured images. When processed in image processing software such as Trimble Business Center software, the PPK data produces high absolute accuracy deliverables.

The UX5 HP aerial imaging rover consists of the wing kit (or body), the eBox, and the gBox. The eBox is a removable box containing the autopilot that controls the UX5 HP. The eBox is connected to a GPS antenna for navigation and a radio antenna for communicating with the ground control station. It is also connected to the camera for sending trigger commands and recording shutter feedback events.

The gBox is a removable box containing a Trimble precision GNSS receiver that provides centimeter-level positioning technology. The gBox is connected to a GNSS antenna for recording positions and to the camera for recording the shutter feedback events.

The UX5 full-frame camera has a large 36.4 megapixel CMOS sensor that provides sharp, detailed images. The standard RGB solution includes a UV HAZE 1 filter. The optional near-infrared solution is for use in specialized applications such as agriculture and includes a B+W 040 (orange) filter. The camera can be fitted with a 15 mm, 25 mm, or 35 mm lens.

The ground control station (GCS) is used to control the UX5 HP aerial imaging rover from the ground. It comprises a GPS-enabled tablet running the Trimble Access™ Aerial Imaging software. The Trimble UX5 HP modem is connected to the tablet to enable radio communication to the rover. The launcher is a mechanical device that provides a safe way to launch the UX5 HP aerial imaging rover in the direction of takeoff. The tracker consists of a transmitter inserted in the body of the UX5 HP aerial imaging rover and a receiver. If required, the receiver is used to track the transmitter signal so the UX5 HP can be located once it has landed.

The typical phases of the UX5 project are

Background map definition—In the office you can prepare the background map, adding details such as avoidance zones.

Block definition—Define one or more blocks to be aerially photographed. You can specify the planned direction the rover will fly over the block(s) in the office, but you must validate the block properties in the field to take into account the actual environment.

Flight definition—Prepare one or more flights to cover the block(s) and suggest the takeoff and landing locations. If the flight preparation is done in the office, then you must validate the flight properties in the field, including the selected takeoff and landing locations.

Flight scheduling—Flight scheduling is completed outside of the Aerial Imaging software.

Complete checks for flight permissions, weather, and site suitability and schedule the flight—Before heading to the site, check all equipment for any damage from previous flights and fully charge all batteries.

PPK operation—If you need centimeter-level absolute accuracy, the UX5 HP should operate with PPK. PPK requires data from a reference station. This station can be a local base station (with or without a precisely known position) that you set up before the flight or a network reference station, such as continuously operating reference stations (CORS), that you access after the flight.

Flight operation—Complete the flight checklist to ensure that the system is ready for the flight. After launching the UX5 HP, monitor the flight using the ground control station. After the UX5 HP lands, complete the postflight checklist to transfer the flight data to the tablet.

Analysis and export—Analyze the flight and captured images to make sure they are synchronized and then export the data in the appropriate file format for image processing.

Postprocessing—Also known as baseline processing. Refine the GNSS data collected during the flight to achieve highly accurate image positions, using postprocessing software such as Trimble Business Center software. The output is called a trajectory.

Image processing—Use the postprocessed data to create high absolute accuracy deliverables such as orthophotos and point clouds, using image processing software such as Trimble Business Center software.

7.5.1 Design Project Plan to Meet Project Requirements

All the factors affecting project performance and accuracy are described in Section 7.3. Current section is devoted to consideration of the specific UX5 HP project planning recommendations. The quality of a project's deliverable is often linked to its accuracy, referring to the positional accuracy of the final output result. A project is processed based on geospatial coordinates, which are automatically acquired during the flight. The output result has a relative accuracy of several times the GSD.

The GSD (see Section 7.2.3) specifies the distance on the ground represented by the pixel of a captured image. It depends on the camera sensor's pixel size and lens focal length. The smaller the GSD value, the more detailed the images. The GSD and the height at which the UX5 HP flies are linked. The higher the rover flies, the larger the distance on the ground represented by each pixel in the images acquired during the flight.

Other factors that impact the final output result include PPK, GCPs, image overlap, and camera settings.

If you need centimeter-level absolute accuracy, the UX5 HP should operate with PPK. PPK requires data from a reference station. This station can be a local base station (with or without a precisely known position) that you set up before the flight or a network reference station, such as CORS, that you access after the flight.

The absolute accuracy can be much improved by using GCPs if PPK is not available. A ground control point is an accurately surveyed coordinate location for a physical feature that can be identified on the ground. Use at least five GCPs, which are evenly spread over the area of interest.

The percentage of image overlap allows processing software to correct for errors in the position and orientation of the aerial images (a rover is a dynamic platform with position and orientation sensors that are not that precise). Trimble recommends having a forward and sideward overlap of 80%, which gives the most accurate results while minimizing flight and processing time. There are some cases where increasing the overlap may be beneficial. Examples are when flying over dense tree canopies, bodies of water, or large areas of sand or snow. These often have a shortage of identifiable features in the images that can be used as tie points for the image adjustment.

Image overlap also varies with changing terrain elevation and features. Use blocks at different heights to ensure the same level of overlap for all areas of interest (Figure 7.27). The image on the left does not take into account the terrain elevation, which results in a varying amount of overlap. The image on the right does take into account the terrain elevation, ensuring the same level of overlap for all areas of interest.

Decreasing the overlap can have an impact on the quality of the orthophoto due to vignetting in the images. This effect can also occur when flying below a certain flight height for a specific lens size or when the texture of the area of interest is homogeneous (for example, bare soil like a desert, short grass, or low growing crops).

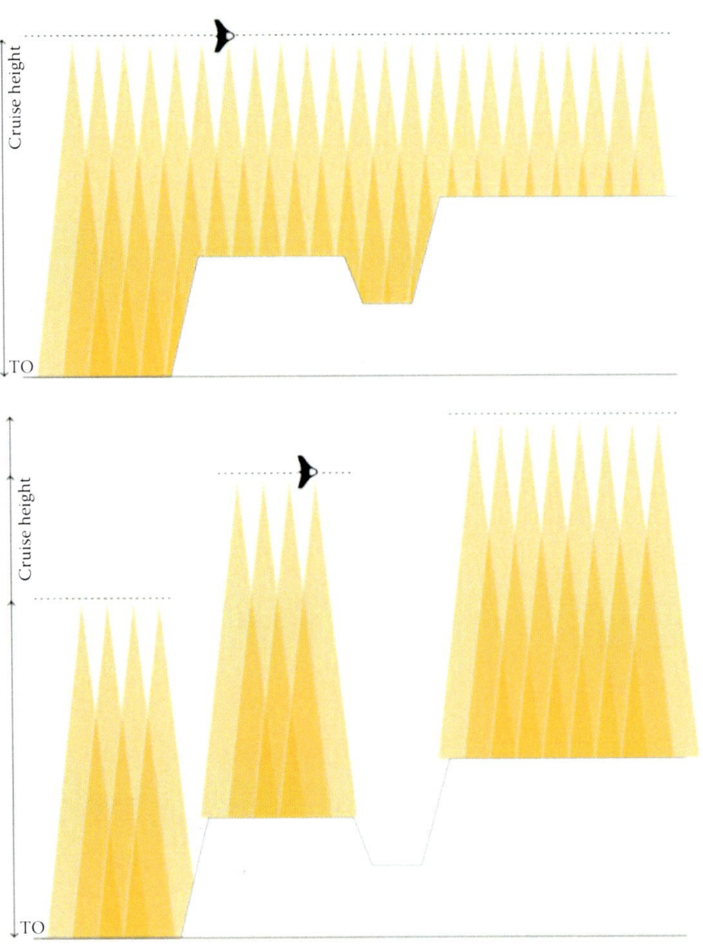

FIGURE 7.27 UX5 HP project planning configuration. (Courtesy of Trimble.)

TABLE 7.6
Field of View/Lens

Lens Size	Resolution at 100 m (328 ft)	Diagonal View Angle
15 mm	3.2 cm (1.3 in)	110.4°
25 mm	2.0 cm (0.8 in)	81.6°
35 mm	1.3 cm (0.5 in)	63.3°

Due to the image perspective, occlusions, or hidden areas in vertical objects may occur, causing undesired distortion effects. Trimble recommends a minimum flight height of 120 m above vertical objects of interest, such as buildings or bridges, in order to avoid distortion in the images.

During image acquisition, the UX5 HP has an average speed of 23.5 m/s. Because of the forward motion of the camera, and depending on the flight height, the shutter speed should be set fast enough to avoid blur effects on the imagery. Therefore, whenever ambient light conditions allow, Trimble recommends not to exceed half of the ground sample distance in forward motion during exposure.

For camera principal focal length considerations, it is necessary to take into account differences in the diagonal field of view summarized in Table 7.6.

Taking into account that the lens size can have an impact on the processed orthophoto due to vignetting issues in the images. Trimble strongly recommends correcting the images for vignetting when using the NIR camera, or when using the RGB camera with a

- 15 mm lens and flying below 100 m (328 ft)
- 25 mm lens and flying below 130 m (427 ft)
- 35 mm lens and flying below 260 m (853 ft)

For the project execution there are two operating modes available:

1. PPK, for centimeter-level absolute accuracy
2. No PPK, for high relative precision

Accuracy specifications according to the ASPRS standard for aerial triangulation (ASPRS 2014) for these operation modes are listed in Table 7.7.

TABLE 7.7
UX5 HP Accuracy Specifications

Coordinates	*X, Y*	*Z*
Absolute accuracy	Down to 2.2 cm (0.88 in)	Down to 2.0 cm (0.79 in)
Relative accuracy	1–2 × GSD	

The absolute accuracy depends on the accuracy of the reference station and the distance between that station and the rover (typically called the baseline length). The shorter the distance, the more accurate the results will be. PPK requires data from a reference station. This station can be a local base station that you set up before the flight or a network reference station, such as CORS, that you access after the flight.

A local base station is a GPS receiver located at a known fixed location or at a location with unknown coordinates. If the coordinates are unknown, then the local base station must be used with data from a reference network, such as the CORS network. A network of reference stations, such as a CORS or VRS (virtual reference station) network, is a network of base stations at a known fixed location. Each base station continuously collects and records GNSS observation data. Many reference networks exist over the world, some of them denser than others. The reference data from a network reference station is accessible via the Internet.

Trimble recommends selecting a network reference station that is GNSS-capable, preferably logs at a rate of 20 Hz (the rate of the rover receiver), and is located within 5 km (3.1 miles) of the area to map. If you do not have a network reference station within that limit, you can set up a local base station that is located within 5 km from the area to map. The distance between the local base station and the network reference station can be much larger than 5 km.

7.5.2 Flight Plan Design

Typically, a UX5 flight is organized as a block or group of blocks. Standard GSDs for that flight planning pattern are summarized in Table 7.8. The creation of a flight plan creation is interactive, and addresses all the issues that might appear such as no fly zones, obstacle avoidance, and clearance for descent. A sample of the finalized UX5 flight plan is depicted in Figure 7.28.

Trimble software allows one to simulate the flight with all the elements and precautions at each waypoint of the flight plan. Flight execution limitations are summarized in Table 7.9.

7.5.3 Completing the Flight

Before flight execution, all the batteries have to be charged and UX5 blocks inspected. Weather and other considerations are described in Section 7.3. It is important to

TABLE 7.8
GSD/Flight Pattern

Height	GSD (15 mm lens)	GSD (25 mm lens)	GSD (35 mm lens)
Minimum: 75 m (246 ft)	2.5 cm (1.0 in)	1.5 cm (0.6 in)	1.0 cm (0.4 in)
100 m (328 ft)	3.2 cm (1.3 in)	2.0 cm (0.8 in)	1.3 cm (0.5 in)
150 m (492 ft)	4.8 cm (1.9 in)	3.0 cm (1.2 in)	2.0 cm (0.8 in)
Maximum: 750 m (2460 ft)	23.9 cm (9.4 in)	15.0 cm (5.9 in)	10.0 cm (3.9 in)

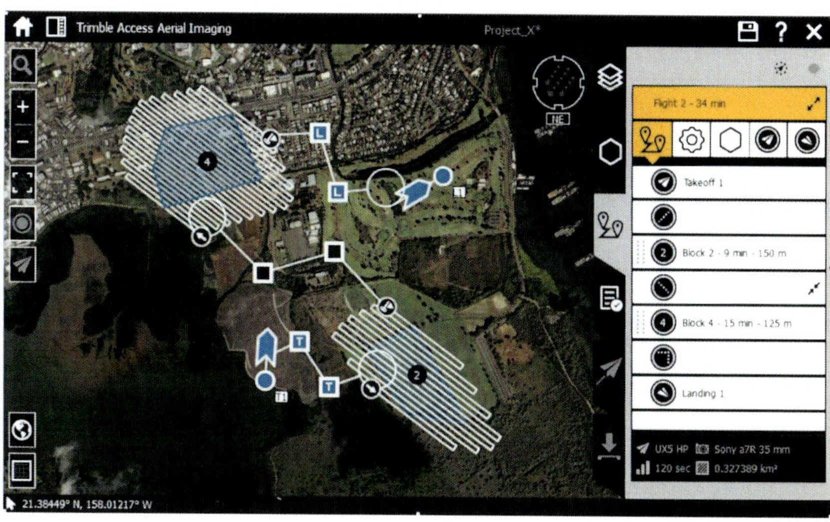

FIGURE 7.28 UX5 HP flight plan. (Courtesy of Trimble.)

make sure that a VLOS (visual contact with rover) is provided for the planned flight project.

The UX5 is launched by means of a cord-based launcher dock, as is depicted in Figure 7.29.

TABLE 7.9
Flight Execution Limitations

Condition	Acceptable Range
Time	Between sunrise and sunset
Distance from clouds	Clear of clouds and always within line of sight
Flight visibility	5000 m (16,400 ft)
Operator–UAS visibility	Visual line of sight
Weather limitations	Light rain is acceptable. Avoid hail, snow, and heavy showers
Head wind (for cruise flight)	Maximum 55 kph (34 mph)
Crosswind	
For takeoff/landing	Maximum 30 kph (19 mph)
For cruise flight	Maximum 55 kph (34 mph)
Gusts (for cruise flight)	Maximum 15 kph (9 mph)
Turbulence	Avoid turbulence at all times
Temperature	
Rover, including gBox and eBox	−20°C to +45°C (−4°F to 113°F)
Camera*	0°C to +30°C (32°F to 86°F)
Battery*	0°C to +30°C (32°F to 86°F)
Launcher*	+10°C to 45°C (50°F to 113°F)

* Preflight conditioned temperature.

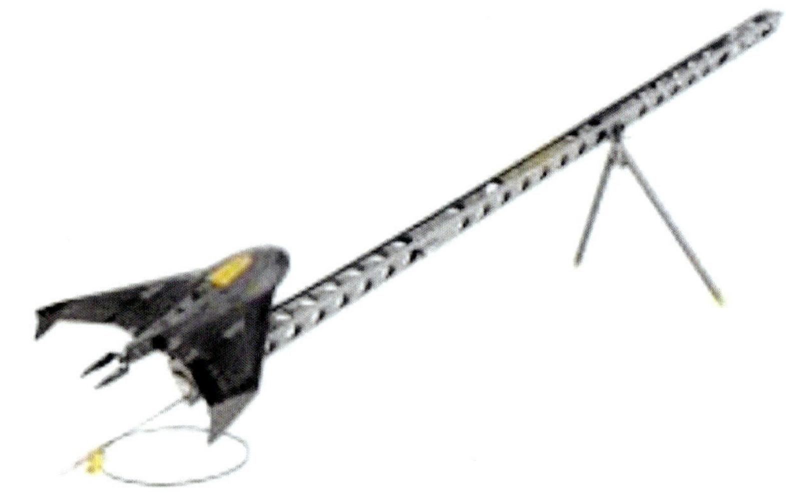

FIGURE 7.29 UX5 on launcher dock. (Courtesy of Trimble.)

After takeoff, the rover climbs to the specified cruise height. Once the rover has passed its first waypoint, the second waypoint is initiated. From this point, the rover follows the preprogrammed flight path. Special Trimble Aerial Imaging software on a tablet can be used during the flight for monitoring and controlling the UX5, as depicted in Figure 7.30.

Specifically Aerial Imaging software allows one to perform the following functions:

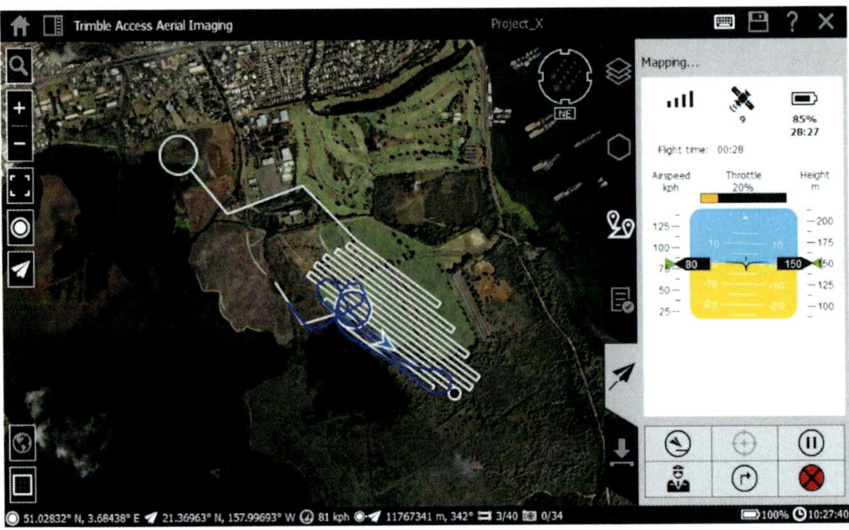

FIGURE 7.30 Monitoring and controlling UX5 during the flight. (Courtesy of Trimble.)

1. Monitor the trajectory of the rover on the map, which must be close to the programmed flight path.
2. Monitor the strength of the communication link.
3. Monitor the number of usable satellites
   - No signal: No GPS lock (0–3 satellites)
   - Signal line: 4–6 satellites
   - Signal lines: 7 or more satellites
4. Monitor the level of power in the UX5 HP battery. In the battery status bar, the value is expressed as a percentage and as time remaining in minutes.
5. Compare the actual flight height with the desired flight height set during block and flight planning.
6. Compare the actual airspeed with the desired airspeed, which is 85 kph (53 mph) during cruise flight. The actual airspeed can be higher than the desired speed when the UX5 HP is coping with a high head wind. The maximum airspeed is 90 kph (56 mph).
7. Monitor the throttle. The throttle should be around 40% during cruise flight in normal conditions.
8. Monitor the software for in-flight warnings and errors such as loss of GPS lock and loss of communication link.
9. Monitor the number of flight lines flown against the total number of flight lines (displayed in the status line below the map).
10. Monitor the number of eBox feedback events against the number of trigger commands (displayed in the status line below the map).

After flight, the next steps are

1. Transferring the images from the camera SD card
2. Returning the project to Aerial Imaging desktop software
3. Exporting flight data for processing
4. Exporting flight path for analysis in Google Earth (KML/GPX formats)
5. Final photogrammetric production within Trimble Business Center Advanced UAS mode

7.5.4 Application Scenarios Supported by UX5 UAV

Per the Trimble UX5 HP manual, the system can be used for

- Orthophoto creation
- Digital elevation modeling (DEM) or digital surface modeling (DSM)
- Photogrammetry

The resulting data is useful for a range of applications, including but not limited to

- Topographical surveying, particularly in remote or difficult-to-access areas, typical in the mining and dredging industries
- Vegetation monitoring
- Infrastructure mapping

REFERENCES

Aber, J. S., I. Marzolff, and J. B. Ries. 2010. *Small-Format Aerial Photography: Principles, Techniques and Geoscience Applications*. Amsterdam, The Netherlands: Elsevier.

Ardu Pilot. 2016. Mission planner overview. http://ardupilot.org/planner/docs/mission-planner-overview.html.

Army UAS CoE Staff. 2016. *U.S. Army Unmanned Aircraft Systems Roadmap 2010–2035*. Fort Rucker; AL: U.S. Army UAS Center of Excellence. http://www.rucker.army.mil/usaace/uas/US%20Army%20UAS%20RoadMap%202010%202035.pdf.

ASPRS. 2014. ASPRS Positional Accuracy Standards for Digital Geospatial Data. *Photogrammetric Engineering & Remote Sensing* 81, no. 3:A1–A26. http://www.asprs.org/a/society/committees/standards/Positional_Accuracy_Standards.pdf.

Blom, J. D. 2010. *Unmanned Aerial Systems: A Historical Perspective*. Fort Leavenworth, KS: Combat Studies Institute Press.

Colomina, I. and P. Molina. 2014. Unmanned aerial systems for photogrammetry and remote sensing: A review. *ISPRS Journal of Photogrammetry and Remote Sensing* 92:79–97.

Colomina, I., M. Blázquez, P. Molina, M. E. Parés, and M. Wis. 2008. Towards a new paradigm for high-resolution low-cost photogrammetry and remote sensing. In *International Archives of the Photogrammetry, Remote Sensing and Spatial Information Sciences, ISPRS Congress*, Beijing, China, XXXVII. Part B1, 1201–1206.

Edward, E. M., J. S. Behthel, and J. Chris McGlone. 2001. *Introduction to Modern Photogrammetry*. New York: Wiley.

Eisenbeiss, H. 2011. The potential of unmanned aerial vehicles for mapping. *Photogrammetric Week* 11: 135–145.

Falkner, E. and D. Morgan. 2002. *Aerial Mapping: Methods and Applications*. Boca Raton, FL: Lewis Publishers.

Goshi, D. S., K. Mai, Y. Liu, and L. Bui. 2012. A millimeter-wave sensor development system for small airborne platforms. In *2012 IEEE Radar Conference 0510–0515*, May 7–11.

Hartley, R. I. and P. Sturm. 1997. Triangulation. *Computer Vision and Image Understanding* 68, no. 2:146–157.

Koo, V. C., Y. K. Chan, G. Vetharatnam, M. Y. Chua, C. H. Lim, C. S. Lim, and M. H. Bin Shahid. 2012. A new unmanned aerial vehicle synthetic aperture radar for environmental monitoring. *Progress In Electromagnetics Research* 122:245–268.

Kraus, K. 2007. *Photogrammetry: Geometry from Images and Laser Scans*. Walter de Gruyter, Berlin.

Kruppa, E. 1913. Zur Ermittlung eines Objektes aus Zwei Perspektiven mit Innerer Orientierung, Sitz.-Ber. Österreichische Akademie der Wissenschaften, Mathematisch-Naturwissenschaftliche Klasse, Sitzungsberichte, Abteilung IIa 122:1939–1948.

Leica Geosystems. 2016. Aibot X6 robust and reliable UAV solution. http://leica-geosystems.com/products/airborne-systems/uav/aibot-x6.

Levin, E., I. Tellidis, and A. Grechishev. 2013. Open photogrammetry. In *Proceedings of the 2013 ASPRS Annual Conference*, Baltimore, Maryland.

Lin, Y., J. Hyyppa, and A. Jaakkola. 2011. Mini-UAV-borne LIDAR for fine-scale mapping. *IEEE Geoscience and Remote Sensing Letters* 8, no. 3:426–430.

Linder, W. 2003. *Digital Photogrammetry: A Practical Course*. Berlin: Springer.

Longuet-Higgins, H. C. 1984. The visual ambiguity of a moving plane. *Proceedings of the Royal Society of London B: Biological Sciences* 223, no. 1231:165–175.

Lowe, D. G. 2004. Distinctive image features from scale-invariant keypoints. *International Journal of Computer Vision* 60, no. 2:91–110.

Lozano, R. 2010. *Unmanned Aerial Vehicles: Embedded Control*. Hoboken, NJ: Wiley.

Luhmann, T., S. Robson, S. Kyle, and I. Harley. 2006. *Close Range Photogrammetry: Principles, Methods and Applications*. Scotland, UK: Whittles.

National Aeronautics and Space Administration, Science Mission Directorate. 2010. Introduction to the electromagnetic spectrum. http://missionscience.nasa.gov/ems/01_intro.html.

Nouwakpo, S. K., M. A. Weltz, and K. McGwire. 2015. Assessing the performance of structure-from-motion photogrammetry and terrestrial LiDAR for reconstructing soil surface microtopography of naturally vegetated plots. *Earth Surface Processes and Landforms* 41, no. 3:308–322.

Ohta, J. 2007. *Smart CMOS Image Sensors and Applications*. Boca Raton, FL: CRC Press.

Paine, D. P. and Kiser, J. D. 2012. *Aerial Photography and Image Interpretation*. Hoboken, NJ: John Wiley & Sons.

Paiva, J. 2011. Will Unmanned Airborne Systems Be the Next Game Changer? *POB*, May 31. http://www.pobonline.com/articles/98734.

RIEGL UAV laser scanning. 2016. 3D Laser Mappping. http://www.3dlasermapping.com/riegl-uav-laser-scanners/.

Sandau, R. 2009. *Digital Airborne Camera: Introduction and Technology*. Dordrecht: Springer.

Topcon. 2015. Topcon UAV solutions for mapping & inspection. http://go.topconpositioning.com/l/74142/2016-02-16/3wglg6.

Triggs, B., P. F. McLauchlan, R. I. Hartley, and A. W. Fitzgibbon. 2000. Bundle adjustment—A modern synthesis. In *Vision Algorithms: Theory and Practice*, edited by B. Triggs, A. Zisserman, and R. Szeliski, 298–372. Berlin: Springer.

Trimble. 2013. *Trimble UX5 HP Aerial Imaging Solution, User's Guide*. Sunnyvale, CA: Trimble.

Trimble UAS. n.d. http://uas.trimble.com/.

Tulldahl, H. M. and H. Larsson. 2014. Lidar on small UAV for 3D mapping. *SPIE Security + Defence*, 925009.

Weishampel, J. F., G. Sung, K. J. Ransom, K. D. LeJeune, and H. H. Shugart. 1994. Forest textural properties from simulated microwave backscatter: the influence of spatial resolution. *Remote Sensing of Environment* 47, no. 2:120–131.

Valavanis, K., P. Oh, and L. A. Piegl, eds. 2008. *Unmanned Aircraft Systems: International Symposium on Unmanned Aerial Vehicles, UAV '08*. Dordrecht: Springer.

Wolf, P. R., and B. A. Dewitt. 2000. *Elements of Photogrammetry with Applications in GIS*. New York: McGraw Hill.

Wolverton, J., II. 2012. U.S. Air Force training more drone, than traditional, 'pilots.' *New American*, August 4. http://www.thenewamerican.com/tech/item/12322-drone-technology-accelerates-usaf-turns-attention-to-training-drone-pilots, archived at http://perma.unl.edu/C57D-GMFV.

Index

A

B

C

D

E

L

M

Q

R

S

T

W

Z